W9-ARR-604

Further Praise for *Oak: The Framework of Civilization*

"This splendid acknowledgment of a natural marvel deserves to be another *Longitude*." —*Publishing News*

"Oak trees have participated in a surprising swath of human history, and now they have finally been recognized for it. William Logan's *Oak* is an utterly fascinating story and, in a strange way, a humbling one."
   —John Berendt, author of *Midnight in the Garden of Good and Evil*

"*Oak: The Frame of Civilization* has a broad appeal, ranging across history, shipbuilding, engineering, forestry and anthropology. But it's a comforting tale as well for those of us with the mighty limbs still dangling over our roofs." —Lynn N. Duke, *Orlando Sentinel*

"A witty, ironic, self-effacing, elegantly crafted—and learned— cultural history of a generalizer in nature: the oak. . . . *Oak* is equally for those who think they know all about trees and for those who never thought twice about them. None will ever contemplate an oak in the same way again."
   —Shepard Krech III, author of *The Ecological Indian*

"[Logan's] underpinning achievement. . . is to make us appreciate just how central one family of trees has been to a whole spectrum of human activities and achievements." —Mark Cocker, *Science*

"With an unabashed enthusiasm for his subject, Logan speaks almost conversationally of the oak's attributes, offering a comprehensive and entertaining history of this highly adaptable and overwhelmingly valuable resource." —Carol Haggas, *Booklist*

"One of the year's most absorbing and thoughtful books in any category. . . . Logan demonstrates persuasively how oaks have shaped who we are and how we got this way."
   —Patricia Jonas, *Plants & Garden News*

"With the luminous clarity and exuberant detail of one who loves what he writes about, Logan traces the ways in which humans have shaped, and in turn been shaped by, the versatile, hospitable oak. By the time you emerge from this engrossing book, you'll be convinced that we are descended from trees. Darwin showed that our remote ancestors climbed down from the branches to stand upright on the ground. Logan shows that our more recent ancestors have used every portion of the oak to meet nearly every human need, from shelter to shoes, from worship to warfare, from filling our stomachs to tracking the stars."　　　—Scott Russell Sanders, author of
Hunting for Hope and *The Force of Spirit*

"*Oak* has plenty to teach us. . . . In his own way, Logan is restoring knowledge that disappeared, and in the end it is his passion for the trees themselves that makes this book remarkable."
—Anthony Doerr, *Boston Globe*

"I don't do jacket blurbs, but I haven't seen a book in years I'd rather write one for than *Oak* if I did. It's a wood-lover's delight."
—Andrew A. Rooney

"Certified arborist and nature writer William Bryant Logan has brought a literary voice to the story of the mighty oak. This wonderful history is written in a storyteller's voice."
—Debra Prinzing, *Seattle Post-Intelligencer*

"Eloquent . . . explained in fascinating detail. The writing style, while very personal and story-based, is packed with both information and insight."　　　—G. D. Dreyer, *Choice*

"A generous and appreciative offering of oak lore."　　　—*Orion*

"An entertaining and instructive homage to the oak."
—*Publishers Weekly*

OAK

The

FRAME

of

CIVILIZATION

William Bryant Logan

W. W. NORTON & COMPANY
New York · *London*

For information about permission to reproduce selections from this book, write to
Permissions, W. W. Norton & Company, Inc., 500 Fifth Avenue, New York, NY 10110

Manufacturing by The Courier Companies, Inc.
Book design by JAM Design
Production manager: Andrew Marasia

Library of Congress Cataloging-in-Publication Data

Logan, William Bryant.
Oak : frame of civilization / William Bryant Logan.
p. cm.
Includes bibliographical references and index.
ISBN 0-393-04773-3 (hardcover)
1. Oak—History. 2. Oak—Social aspects—History. I. Title.
SD397.O12L64 2005
634.9'721—dc22

2004028385

ISBN-13: 978-0-393-32778-6 pbk.
ISBN-10: 0-393-32778-7 pbk.

W. W. Norton & Company, Inc., 500 Fifth Avenue, New York, N.Y. 10110
www.wwnorton.com

W. W. Norton & Company Ltd., Castle House, 75/76 Wells Street, London W1T 3QT

3 4 5 6 7 8 9 0

*For Bear
and Nora*

Not unfortunately the universe is wild. . . .

WILLIAM JAMES

It's not what we see, but what we see in it.

EDGAR ANDERSON
TO JOHN BRINCKERHOFF JACKSON

CONTENTS

ACKNOWLEDGMENTS

I AM NEITHER a botanist nor a historian. This book could not have been written without the help of many generous people and scholars from around the world of oak.

Let me first of all thank Professor Kevin Nixon of Cornell University, whose insights on the "greatness" of oak were crucial to focusing the book. I am also deeply indebted to Ole Crumlin-Pedersen and his staff at the Viking Ship Museum in Roskilde, Denmark, who patiently taught me about Viking ships, as did Arne-Emil Christensen at the Viking Ship Museum in Oslo, Norway.

For help on the balanoculture section, I most gratefully acknowledge Sarah R. Mason, whose dissertation on acorns as human food provided a roadmap for my research. Likewise, I am indebted to David Bainbridge, who so far as I know coined the word *balanoculture* and whose thesis about acorn eating helped shape my own. Suellen Ocean was generous with her time and her recipes, both of which helped me learn to make acorn dishes for myself. Delfina Martinez of Hopland, California, welcomed a perfect stranger and answered his numerous intimate questions about her childhood.

My research in England was wonderfully and unexpectedly helped by Brian Roby, the proprietor of Felton House, Hereford, who sent me all over the countryside in search of trees, timber-framed houses, and craftsmen whom he knew. Steve Potter, forester, showed me the forest from which the new roof for Windsor Great Hall had come. Mark Hicks of Capps & Capps Ltd. discussed their work on that project and timber building in general. He walked through his woodyard, commenting on the uses and qualities of each piece. John Green of Border Oak, Herefordshire, walked me through the whole process of timber-framing, using his company's own products as examples, and Julian Monkley, cabinet maker, showed me through the timber-framed house he'd built himself.

Other craftsmen were equally generous with their time and knowledge. Mike Greason, consulting forester, walked me through the snowy woods, evaluating oaks and explaining their insides by looking at their outsides. David Proulx, the longtime cooper in residence at Sturbridge Village, showed me the steps in caskmaking and the careful judgment required to get it right.

Dwight Demilt, the carpenter in charge at the USS *Constitution,* was very generous with his time and expertise. He led me from stem to stern and top to bottom of the ship, explaining everything as he went. Librarian Kate Lennon-Walker at the *Constitution* Museum was a great help in finding references not only to the ship itself but also to many matters related to sailing and sailing ships in general.

The tree physicist Klaus Mattheck was also important to me, not only as a general teacher about the ways of trees but also for his pioneering ideas about using tree structure as a model for better industry.

So many people helped shape this book. I also thank Ray Tanner, now deceased, and Bob Lansing for their comments on the woods and town where they were raised. Rabbi and Mrs. Kornbluth were an education to me about real respect for the natural world.

Many thanks too to Wayne Cahilly and to David Sassoon, who read the manuscript critically.

There would have been no book without John Barstow, who asked me to write it and who never lost interest through many delays, or without Alane Mason, my editor, and Alessandra Bastagli, her assistant. Alane insisted that the book be as good as it could be and tirelessly supported it.

My deepest thanks also go to my wife, Nora H. Logan, whose beautiful illustrations are an important part of the book. She and our children, Sam, Jake and Eliza, were also endlessly patient when Dad disappeared into the trees yet again. Finally, I am indebted to the men of Urban Arborists, especially to Fernando, who have been my constant companions among the trees for more than a decade.

OAK

The Frame of Civilization

In the Middle of
the World

Trees are the tallest, most massive, and longest-lived creatures on earth. But oaks hold none of the records. The oak is not the tallest tree. That distinction belongs to a redwood in northern California, which tops out at about 368 feet, easily twice as tall as the tallest oak. Neither is the oak the most massive tree. That honor is held, if you wish, by the General Sherman sequoia in Kings Canyon National Park, weighing in at about two thousand tons, though if you are thinking genetically, the record may instead be held by a clonal aspen grove in Utah that covers more than 106 acres. The oak is not the oldest tree. The bristlecone pine holds that record, at something in excess of 4,867 years. The oak is not the strongest tree. Ebony, teak, and many other tropical woods have greater strength under tension, compression, and shearing forces. And the oak is far from the fastest-growing tree. There is a species of albizia in Malaysia that can grow more than an inch each day; an oak is lucky to do a foot each year. Any old maple, pine, poplar, or eucalyptus can easily outrun the oak.

So what is so special about oaks?

I asked Kevin Nixon, a paleobotanist at Cornell University, who studies the origin of oaks.

"Nothing," he said, and paused.

"But what is impressive about them is that you can go from Massachusetts to Mexico City and find that the same genus—the oaks, that is, *Quercus*—is dominant, when there are very few other genera that are even common to both places."

"Well, why is that?" I asked, trusting that oak held some world record at least.

"No reason," he replied nonchalantly.

Quercus hindsii, California white oak (Collection of William Bryant Logan)

It was a good thing we were speaking by phone, since he did not see my cheeks puff out and redden as I tried to keep from spluttering. There was an awkward silence.

"It's like the chambered nautilus," he continued, at last. "The *Nautilus* genus was once very diverse, but then it over-specialized until it could only live in one particular niche in one particular way."

"Go on," I said. He had my interest.

"But the oaks never overspecialized," he continued. "They never found a niche. They are so successful exactly because there is no reason that they are. Restricted distribution only happens when there is just one reason for a creature's success."

Here was an unexpected and tantalizing idea. The persistent, the common, the various, the adaptable has value in itself. The oak's distinction is its insistence and its flexibility. The tree helps and is helped in turn. It specializes in not specializing.

And indeed, the champion trees are niche holders: The giant redwood can only survive on a cool, fogbound band of warm coast. The ancient bristlecone pine only grows on one high mountain ridge, where no pest can survive long enough to attack it. Ebony needs great heat, great water, and little varia-tion in temperature to make its strong, flexible, durable wood. The clonal aspen thrives and expands only at altitudes and on exposures high and difficult enough to ward off competing species. None of these record trees thrives throughout the tem-perate zone, in the middle of the world, where the unexpected is commonplace.

Names and Forms

I WALK IN woods that once were fields. When I leave the parking lot in the limestone hills, a thousand feet above the Hudson Valley, I step instantly into a world of trunks and branches. Most of the trees are small. There are the pioneer beeches that spread by root sprouts, so that a thousand little beeches are but one tree; the young tough ironwood and hop hornbeam; the basswood with its big leaves on which all kinds of bugs love to light; the sugar maple and in the bottoms the red maple; and the striped maple, its green young stems streaked with white like some stiff upright grass snake; the amelanchier, also called serviceberry, because its tasty fruits set in spring just about the time that the ground thaws enough to bury the dead; the hemlocks that survived mass cuttings in the nineteenth century and that now eke out a living in the shade of other trees, growing perhaps a sixteenth of an inch per year, waiting for a change of days; and the spruces planted in blocks during the 1930s by the Civilian Conservation Corps and now standing in ragged funereal groves. On the ridgelines where the sun is best stands the northern red oak, *Quercus rubra*, branches twisting for the light, roots impossibly jointed finding their way downhill over sheer cliffs and through the fissures in stone, bark mottled with half a dozen species of lichen.

The biggest trees in this forest are the oaks. The broadest are maybe twenty-four inches in diameter at the height where I can embrace them. When all the other trees were cut down to make room for farm fields, some of the oaks were left to mark the boundaries.

Ray Tanner, who grew up in these parts through the whole course of the twentieth century, thinks the world of his old age is ass-backwards. When he was a kid, he remembers, the trees were in town, and all the hills were planted to hay or corn or wheat. "You could see around you then," he said. "It wasn't so closed in." If you wanted to gather hickory nuts or butternuts, you had to know along which hedgerows to look. When winter came, people put away the wagons and took out the sleighs. Kids sledded down the main street of town under the branches of sugar maples and elms.

Dutch elm disease took the elms. When cars came, the townships started to salt the roads to melt the ice, and the salt killed off the sugar maples. There are still a few survivors, stag-headed and with a large column of rot almost bisecting each tree between the main leader and the subleader, the stems into which almost all sugar maples divide. I never park my car beneath them. When I see a car there I sometimes imagine that the tree is about to exact its revenge for the profligate century that killed it.

One by one, the farms went out of business. Bob Lansing—he and Ray were best men at each other's weddings—used to be the local feed salesman. He remembers that early in his career he'd had dozens of customers in the hill country. At the beginning of the 1970s he still had eighteen. By the late 1970s, only one was left. As the farms declined—a process that began early in the twentieth century—the woods moved in. No unplowed field in this part of the world stays open for long. A good forester can describe the agricultural history of the region by looking at the different ages of woodland stands.

Some landowners cut hardwoods in the forest stands and make a decent profit. In fact, the state of New York has become one of the leading sources of hardwood lumber in the world, owing entirely to these former farm fields. Otherwise, the woods go unused, or they are used only by great nature.

Humans pump more and more carbon dioxide into the air by burning fossil fuels, but according to many scientists, there ought to be a lot more CO_2 than there actually is. Where is the rest of it? Many think it is going into the regenerating woodlands, that the carbon dioxide that might harm us in the air is in fact becoming the solid stuff—the long-chain carbon molecules, including proteins and amino acids—in the trees.

Still, the average person might look at these woods and think, "I could clear away here and put up my house," or "If I want to make a garden, which trees shall I keep, or shall I cut them all so I get more light?" or "We'll grub out ten acres, take it down to subsoil, compact it, and build a mall." The main question about the trees is the cost to dispose of them. Hardly anyone thinks that wood once constituted the greatest part of the wealth of tribes and of nations.

Since the glaciers last retreated and since humans began to build and settle down, there have been but two versions of the world: the world made with wood and the world made with coal and oil. One lasted twelve to fifteen millennia; the other has lasted about 250 years so far.

All that it is to be human was defined in and through wood: the house and the town, the wagon and the plow, the ship and the shirt, the post office and the dancing ground, the pen and the window, the bath and the cask, the wine bottle and the goblet, the realms of gods and devils, the symbols of fertility and death. The world of coal and oil has only adapted, or parodied, them. These later fuels, indeed, embody no new principle. They are just old stem and branch and leaf and root, entombed for eons and distilled, the remnants of a wooden world from before the coming of humankind.

In most of the temperate world, oak is the primary, the titular tree of the forest. In Sanskrit, the name for the oak and the name for trees in general are the same: *duir*. No tree has been more useful to human beings than the oak. It was the oak that taught humans forestry. Its composition made it the easiest wood to split and shape. A drying log showed the internal network of circles and radii along which it could be split. One could make planks or beams of any width or thickness, limited only by the size of the tree. Oak alone could so flexibly and reliably be shaped with stone axes. Once bronze and then iron tools appeared, oak became a chief strategic material.

To discover the world that made us, look at what it has left us—half-timbered houses; Leonardo da Vinci's drawings executed in oak gall ink; Viking Age oaken ships buried with the dead; Bronze Age oak log coffins; ancient barrels, casks, vats, and tubs; wine corks and truffles; fossil leaves from thirty million years ago that just might be from the first oaks; layers and layers of what botanists call "pollen rain"; living oaks over five hundred years old and the black hulks of oaks that drowned beneath the rising seas ten thousand years ago, trees that don't have even a branch until ninety feet up the trunk.

But to begin with, just look at the artifacts of names.

Embedded in the names of people and places we meet every day is the memory of the woods that shaped the lives of our ancestors. Many families are named directly for forests: the Woods and Atwoods and Forests; or their French cousins, the Bois, Dubois, and Dubos; or their Italian and Spanish cohorts, the Boscos, delBoscos, and Buscos; or the German Walds, Wildes, Waldbaums, and Holsts. There are people named for some place in the woods: the Bradleys of the broad glade; the Brants, Bryants, Brandons, and Brandos of the place cleared by fire; the Shaws of the thicket; the Dewhursts of the wooded hill; the Brentwoods of the burnt grove; the Greves of the thickets and the Henleys of the clearing of the high woods; the Reeds and Ryders of the clearings; the Chathams of the forest homesteads; the Waylunds and the Lunds of the sacred grove; the Sylvesters, the woodland ones; the Chases of the hunting park; and the Mangroves of the woods that are part of the commons. And the people named for some type of woods: the Spinneys of the thorn wood, the Birchetts of the birch wood, the

Carrs of the alder wood, the Tellers of the large oaks reserved for the king, the Carpinatos of the hornbeam woods, the Hollingsworths of the holly woods, the Hasslers of the hazel grove, the Hawthorns and Hagedorns of the hawthorn woods, the Hesters of the beech wood, and the Cashes of the oaks.

Even more common are the names derived from what people did in the woods. All the men and women first called Johnson and Robinson, as well as Hobson, Dobson, and Everson, were named because no one knew who their fathers were. It was the custom in the Middle Ages and before to go into the woods on Midsummer's Eve for revelry that included a great deal of promiscuous coupling. No one thought this wrong, but the children of such unions would be called Johnson, son of St. John, since Midsummer's Day is on St. John's Day; Robinson, son of Robin Goodfellow, whose other names were Hob and Dobbie; or Everson, son of Eve.

If some extraterrestrial race were to find no more of humanity than records kept on some future colony far from Earth, if they could reconstruct the original meanings of the names, they could learn how people lived. It is a history in which wood and the greenwoods play the leading part. Between the thinking and the making, a world came into existence and hominids began to become human.

The Fosters and Foresters, the Woodwards, Wards, and Haywards distributed the rights to use certain trees in certain ways. The Hoggs pastured their swine on fallen acorns. The Cobbs pollarded the trees, cutting them back to between eight and fifteen feet tall, high enough to be out of the reach of browsing animals. The Hughes and the Fellers dropped whole trees. The

Cleavers, Clovers, and Clevengers split the wood; the Sawyers and the Pitmans sawed it. Then the others went to work. The Barkers harvested the tanbark, and the Tanners cured leather with it. The Coopers, the Hoopers, the Beckers, and the Benders worked the raw wood into barrel staves. The Wheelwrights and the Axelrods turned the wood for spokes, bent it for rims, and shaped large balks into axles. The Carpenters, Turners, and Woodwrights built the half-timbered houses, split the flooring, shaped the joists, turned the furniture, and planed the wood for cabinets. The Boatwrights molded the keels and fit the clenched-lap planks to them. The Crews worked slender strips of wood into weirs for fish traps, while the Cases fashioned boxes. The Colliers, Colemans, and Charbonnels reduced waste wood to charcoal, so that the Smiths, the Fabers, the Febvres, the Levebres, and the Goughs could forge iron and the Paines make glass. The Ploughwrights and the Coulters joined sturdy shaped beams to make drawing shafts for the plows.

Among all lists of names derived from wood, none is longer than the list of those derived from oaks. All names that contain the roots *ac, ech, ag, og, hick, heck, eiche, chene, cas, daru, dru,* and *rove* owe their origin to oak. The Aikmans, the Eichengreens, the Eichorns, the Actons, Akroyds, Akbars, Oakhams, Wokinghams, Oakleys, del Encinas, Encinals, della Roveres, Chaneys, and Cashes are all of oak. It is the most widely used tree name in all the Western languages, from Sanskrit to Celtic, from the Indian subcontinent to the tip of Ireland. The name of the historic Buddha, *Shakyamuni,* means "the sage of the oak tree people."

A prominent Italian family, the della Rovere (of the oak), produced Pope Julius II, patron and probably lover of Michelangelo. To honor his patron, the artist decorated the Sistine Chapel's ceiling with a number of *ignudi*, naked male figures, each reclining on an outsized cluster of acorns. The resemblance between the acorns and the penises of the figures is quite close.

Acorns in the Sistine Chapel ceiling
(Michelangelo, Scala/Art Resource, New York)

Names, names, names. I could go on multiplying them. There are plenty more. But I am in the position of the excavator of an ancient ship, who sometimes will find only an oarlock and have to infer the gunwale, or only a fragment of keel and have to infer the prow, or only an earring and have to infer the man.

When I began this book, one of the first things I wanted to do was to find a map of the worldwide distribution of oaks. I asked a young man who was helping me in the office to get me such a map. It was hard to find, he said. When at last he presented a copy of the map to me, I had my doubts. It appeared that it was a map of the trade routes linking the historical civilizations of the Eastern and Western worlds. He had had to reduce the original map greatly to get it on the photocopy, so the legend was quite small. I squinted. The tiny legend WORLD OAK DISTRIBUTION swam into view. I rubbed my eyes. The caption still read the same.

This map showed clearly that the distribution of oak trees is coterminous with the locations of the settled civilizations of Asia, Europe, and North America. It would be rash to suggest that oak trees were a condition for these civilizations, but it is interesting to think that from Kyoto to Beijing, from Kashmir to Jerusalem, from Istanbul to Moscow, from Gibraltar to Oslo, from New York to Chichen Itza, from Mexico City to Seattle, where there are or have been the cities and cultures that shaped the modern world, there are or have been oaks.

World Oak Distribution (Nora H. Logan)

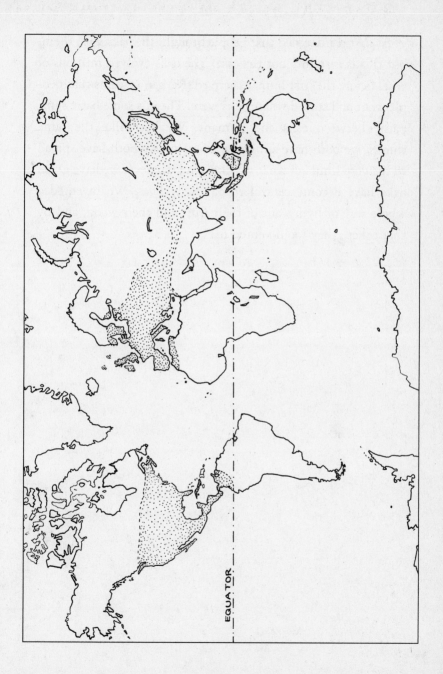

EQUATOR

At first, I imagined that people brought the oaks with them, but this is certainly not the case. The oaks evolved millions of years before the first humans arrived. Rather, it seems that people went and stayed where oaks were. There is some basic sympathy between oaks and humans. We both like the same things, we both have similar virtues, and we both have spread to the very limits of what we like. And wherever we have gone, oaks have become central to our daily lives. We invented a whole way of living out of their fruit and their wood, and by that token, they too invented us.

BALANOCULTURE

WHAT IS A FRUIT? What is a nut? A dessert? A snack food?

A short time ago the Kornbluths, a young Orthodox Jewish couple living in Brooklyn, called me in my capacity as arborist to come and look at their pear tree.

"We need to know if it can bear fruit," Mrs. Kornbluth said over the phone.

I asked her why she thought the tree couldn't bear fruit. She said that an arborist had looked at it recently, told them it was full of carpenter ants, dying, and should be removed immediately. No, he had told them, it would never bear fruit again.

"It sounds awful," I responded. "Why not just remove it, as the arborist suggested? Why call me?"

"We want a second opinion," she said. "You see, under our law, you cannot cut down a fruit tree if it can still bear fruit."

We talked for a long time, Mrs. Kornbluth and I. She had many questions. She wanted to know if I could positively, definitively tell them that it would *never* bear fruit again. I was envisioning a tree three-quarters gone with root decay, dieback on all the stems, and with a very short time to live. I said I could not say with 100 percent certainty, so long as the tree lived—only God could do that—but I could give a reasonable probability. This was all very important, she explained to me, because they wanted to build an addition to the house that they had just purchased. If the tree could bear fruit, they could not proceed. She asked, and I told her my consulting fee. She said she'd talk to her husband and their rabbi and would get back to me. Once she'd hung up, I wondered what kind of culture would show such apparently exaggerated respect for a single fruit tree, and why.

The next day, she called to set up a meeting at the tree with her and her husband. I arrived twenty minutes early and walked around behind the house. I was nervous, anticipating that this would be a very hard call. I wanted to get a head start on the tree, without anxious watchers hanging over my shoulder.

My mouth fell open. I did not see a dying pear full of ants and root decay. I saw a vigorous pear, squeezed in a corner though it was, with leaders jumping up above the roof of the house. It was full to the brim with fat flower buds—the month was January—and almost every growing tip showed large lateral scars down the stem behind it, indicating where a fruit had been.

I was astounded by the difference between the story that the Kornbluths had been told and the story I had to tell them. I doubted the evidence of my own eyes, since I could not believe that the other story—in which I had implicitly come to believe—could be wrong. I had been preparing to try my best to assess the chances of a dying tree, and suddenly I had to tell them about a healthy one. What is more, I knew how Mrs. Kornbluth wanted the story to end. She wanted to build her addition. But she did not want to break Jewish law.

When they appeared, I was disarmed. They were slender, earnest, intelligent, and very young. They had a toddler in a stroller and a babe in arms. She had dark almond eyes and a long green winter coat. He was dressed in black from head to foot, with the fine, black, broad-brimmed hat that the Orthodox wear and a fresh, blooming, reddish beard.

I said what I had to say: the tree was almost certain to bear fruit again, barring an act of God, a bolt of lightning, or the destruction of every other pear tree within five miles that might pollinate it. (Pears are not self-fruitful; they need another pear in order to bear fruit.)

I pruned part of a young branch to show them what I meant. They were chagrined. Had I hurt the tree? No, I said, pruning is good for trees, if it is done judiciously and correctly. They were amazed. They decided to ask their rabbi whether they ought to prune the tree instead of removing it.

Through all of this, I felt bad. I could see the house they'd bought and their two children. I had no idea how many kids they wanted, but you could see that they were delighted to have a family. The house was not large. An addition would have seemed just the right thing to continue their story.

"I'm sorry I don't have better news," I said.

The husband—he must have been ten years my junior—looked at me and with only the slightest vehemence said, "This isn't bad news. This is very good news. Because you have prevented us from doing a thing that is wrong. It is good news, and I am very happy about it."

This brief speech impressed me deeply. He was not being self-righteous. He didn't seem to be trying to win a point against his wife, either. In fact, he appeared to be trying to enjoy obeying an ancient law, even though it might make his life less stable and less ordered. I had known numerous clients and would-be clients quite willing to remove bearing fruit trees and huge centenarian oaks simply because the leaves were littering their yards and clogging their gutters. They were ready to kill a creature more than one hundred years old in order to spend one day less each year sweeping leaves. This man needed to make a bedroom for a child, but he wouldn't kill this tree to do it.

"It's a good law," I said. Neither of them replied. But a strange law, I thought to myself. Where had it come from?

On the way back down the driveway to the street, I started to identify other plants on their property, partly to console them, I suppose, partly because I was nervous in front of them, and partly because I felt that I had hardly earned my consulting fee. It didn't take a genius to know that that tree would fruit. When we got to the small front garden, I showed them a euonymus shrub that had the usual scale infestation, and a pair of lovely upright junipers. "The berries of these junipers," I remarked in passing, "are not eaten anymore, but are sometimes used to flavor gin."

"Ah," they replied. They asked how to treat the scale and about the identities of other plants on the property line. We parted.

Three days later, I got another call from Mrs. Kornbluth.

"What was that you said about jam?" she asked.

I was nonplussed. "I didn't say anything at all about jam," I answered.

"You said one of the plants in front was used for jam. Our rabbi says that that makes it a fruit tree, and we cannot cut it down."

Uh, oh, I thought. I must have put my foot in it again.

"I don't understand," I confessed.

"Since we can't build the addition in back," she explained, "we decided to build it in front, but if those trees are used for making jam, we cannot do that either!"

"Mrs. Kornbluth," I said, "no tree I saw at your house, except the pear, could be used in making jam."

"But you said so clearly," she insisted.

The light dawned in my memory. "Gin!" I exclaimed. "Not 'jam,' I said 'gin.' The berries of the juniper are used in making gin."

"The drink?" she asked.

"The liquor, yes," I said.

"I hear," she said. "I will have to ask the rabbi. May I call you again?"

"Of course," I replied. I saw that I was indeed going to earn my fee.

The next day, I heard from Mrs. Kornbluth again.

"Can you move those trees, the gin ones?" she asked.

"Yes," I said, "in the right season, of course I can. They are not very large, and they are easy to move."

"Will they grow and be healthy?"

"God willing, yes," I said. "They should be fine."

"Good," she said. There was deep relief in her voice. At last, the young mother was going to get her house in order.

"But if you don't mind my asking," I went on, "no one eats the juniper berry, at least not anymore. The Apaches may have eaten it once, but now it is used only to flavor gin. Does that make it a fruit tree?"

"Yes," she said. "The rabbi explained it like this: We use the quercitron in our festivals. It is very bitter, and yet we prepare it and use it sparingly to flavor. So it is a fruit. And even things like oats or wheat, these are changed and prepared before we use them, and parts of them are thrown away."

Just so. Through the Kornbluths, I began to see deeper into time. The juniper berry had indeed been a food to the Apache and to many others. The law of the Kornbluths was more than three thousand years old and might preserve in its pages the memory of a time qualitatively different from ours. I have read the book of Genesis since I was a child, but now for the first time it occurred to me that God had told Adam and Eve to eat the fruit of every tree in the garden, except for one.

To eat the fruit of the trees. Not for dessert. Not as a snack food. To live on the fruit of trees.

What if this were not a poetic manner of speaking, but a description of what people, before agriculture, really did? I had loved the acorn-eating cultures of the Indians of California, where I grew up, but always considered the eating of acorns a habit peculiar to these small, rotund people on the edge of the earth.

What if they were not the only, but only the last of, the balanophages?

Balanophages: acorn eaters.

The Old Stories

Just as we have legends of an Arcadian past, in which pastoral peoples lived on the edge of the Mediterranean, so those pastoral peoples themselves had legends of a past in which people ate the fruit of oak trees.

There is a whole class of old stories that speak of such ancient, prehistoric systems. The question is, were these stories largely made up, or have they more than a grain of truth? To begin with, remember that Troy and Mycenae were thought to be baseless legends until some intrepid archaeologists unearthed them. And though in the nature of things it is impossible to unearth a balanoculture—that is, a culture that lived on nuts— still we may find the story compelling in its truth.

In the eighth century B.C., the Greek poet Hesiod, in his *Works and Days,* asserted that acorns effectively prevented hunger: "Honest people do not suffer from famine, since the gods give them abundant subsistence: acorn-bearing oaks, honey, and sheep."

The poet Ovid repeated the acorn story in *Fasti,* the work of his maturity, which was meant to be a poetic retelling of the feasts of the Roman calendar. Under the feast of Ceres, goddess of grain and agriculture, he recounted that prior to the age of agriculture, people lived on acorns. "The sturdy oak afforded a splendid affluence." Ovid was a contemporary of Virgil and Horace, and their equal as a poet. He was certainly not bandying words about lightly. When he wrote "affluence," he did not just mean that acorns were yummy, but that they dependably provided people with plenty to eat.

Lucretius repeated more or less the same tale, with the addition that oak was so important to people's lives that acorn-decked oak boughs were carried in procession in rites of the Eleusinian Mysteries.

Pliny, the most audacious and voracious story collector of his time, described all the different oaks of which he was aware and their many uses. "Acorns at this very day," he wrote, "constitute the wealth of many races, even when they are enjoying peace. Moreover also, when there is a scarcity of corn they are dried and ground into flour which is kneaded to make bread; beside this, at the present day also in the Spanish provinces a place is found for acorns in the second course at table."

Pausanias, in his *Description of Greece,* written about the year A.D. 160, described the founding of the kingdom of Arcadia by Pelasgus, who, he reports, invented the use of houses, the wearing of sheepskins, and the eating of acorns. An interesting trio: home, clothing, food. It sounds very much like the foundation of what we would call a human life.

Pausanias writes, "He too it was who checked the habit of eating green leaves, grasses, and roots always inedible and sometimes poisonous. But he introduced as food the nuts of trees, not those of all trees but only the acorns of the edible

oak." The suggestion is that, by creating a staple food, Pelasgus was able to create a sedentary life for his people, perhaps
the first of its kind. Though Pausanias was writing about what
to him was antiquity, he notes that still in his own time, the
Arcadians were fond of acorns.

To my mind, however, one of the strongest bits of evidence
is not even a story. It is just a word. The old word for oak in
Tunisia means "the meal-bearing tree."

Later writers in the Latin west picked up the antique stories
and passed them on, at least for a period of time. The English
Renaissance poet Spenser put the matter clearly and beautifully
in his pastoral "Virgil's Gnat":

> *The oke, whose Acornes were our foode before*
> *That Ceres seede of mortall men were knowne,*
> *Which first Triptoleme taught how to be sowne.*

Recently, however, we have come to think these early writers credulous. Everything written before the Scientific Revolution smacks of hearsay. After all, Pliny contended among
numerous other unsupportable assertions that buzzards were
impregnated by the wind. How then can we trust him and his
confreres and their progeny on a matter so important as the origins of human eating habits and of settled life itself?

No, the modern scientific thinkers knew the truth, and the
truth was red in tooth and claw: Pre-agricultural man had been
a hunter. Only when he'd exterminated most of the game and
when a change in climate further reduced his hunting stock of
large quadrupeds did he change to shepherding and agriculture.

Nuts? Oh, prehistoric man sometimes ate them, just as we do today.

In fact, early "scientific" prehistorians remade ancient man in the image of an industrial capitalist. Of course they sought the largest game, and of course they destroyed as much of it as they could. And of course they would have gone on destroying it until some worldwide change forced them to alter their habits, that is, to innovate, to create a new way of life. *That* is the march of progress, *that* is the way things are.

But the researchers who followed these bold theorists began to find unexpected difficulties with that hypothesis. In fact, people had evidently eaten not just large game animals, but rather anything they could lay their hands on, from turtles to snails to eggs to acorns. Furthermore, there never had been any change in climate such as had been proposed. Third, in archaeological finds, there appeared to be grinding tools before there was any evidence that even wild wheat had been cut for human consumption.

What were people grinding?

Perhaps the book on which the Kornbluths' faith is based can give some answers. Genesis has a good deal to say about fruit trees. In the first creation story, which comprises the first chapter of the book, God concludes, "See, I give you every seed-bearing plant all over the earth, and every tree that has seed-bearing fruit on it to be your food." There is no mention of animals as food.

Chapter two of Genesis contains a second creation story, thought to have been composed before the first. This is the tale of the Garden of Eden: "Out of the ground, the Lord God made to grow various trees that were delightful to look at and good for food, with the tree of life in the middle of the garden." God tells Adam, "You are free to eat from any of the trees of the garden, except the tree of the knowledge of good and evil."

In our curiosity about the identity of the Tree of Good and Evil, we lose sight of the main point: that God tells unfallen Man to eat the fruit of trees. What fruit trees were available in Mesopotamia's Fertile Crescent after the end of the last ice age? Oaks, junipers, pistachios, maples, and wild pears. (Yes, Mrs. Kornbluth! Pears.) Of these, all were likely eaten, but the one that had the proper nutritional characteristics to become a staple food was the acorn.

How far that time seems from this! Oaks have been cut down without a thought around the world for three thousand years, though not, I suppose, in any place where the Kornbluths' rabbi or his ancestors and teachers were to be found. But they are not forgotten as a source of foodstuff everywhere or by everyone.

Recently, I took a walk in the heart of New York City's Little Korea on Thirty-second Street in midtown Manhattan. Someone had told me that Koreans still eat acorn products, and that if I looked in a good Korean supermarket, I might still find something.

In the middle of the block was the biggest Korean market in the city, and behind the cash register was a beautiful girl with long straight black hair and a face shaped like a slender acorn of the white oak. I asked her if she knew if there were any foods made from acorns in her store, and she looked mystified. She queried her neighbor at the next register—"Hawgon?"— and she too shrugged her shoulders.

Unwilling to go away empty-handed, I rummaged around in my change pocket, where I usually keep at least one acorn. Sure enough, there was a red oak acorn that I'd collected a few months earlier in New Hampshire. The girls were by this time looking at me dubiously and nervously.

I drew out the acorn and held it up. This, I said, and pointed.

Immediately, the two broke into smiles and laughter. "Of course, of course," said the black-eyed one. "Yes, we will show you." The acorn was obviously as familiar to them as a pinto bean or a rice grain, though not by the name *acorn*. They led me back to the flours, where they picked out a pound of acorn starch flour for $5.99, and then to the cold case, where there was a tofulike square of acorn jelly for $3.99. All the while, one of them kept trying to explain to me how to prepare the jelly from the powder.

Perhaps Pliny and Ovid and Pausanias and all the others were not wrong. Maybe whole cultures once ate the fruit of the oak. If we are children of the Northern Hemisphere, our ultimate fathers and mothers likely fed on acorns.

The Meal-Bearing Tree

A few kids in New England and the upper Midwest still know that if you find a fresh "oak apple" on a red, scarlet, or black oak, you can poke a hole in it and suck out the sweet juice. Most have probably been cautioned against such a risky and unsanitary practice, but even to this day the Kurds, Iranians, and Iraqis swallow the sweet exudate from the oak tree, which they call "manna." Both juice and manna are the distillations of oak sap.

Each June and July, the Kurds wait for the sweet drops to begin to congeal on the leaves of *Quercus infectoria*. Early in the morning, before the ants have had a chance to visit the leaves, they beat the branches over a cloth laid out on the ground, dislodging the crystallized manna. Sometimes it is brownish, sometimes greenish, sometimes as black as tar, and sometimes pure white, but always it is very sweet. They use it to make a breakfast drink, or they mix it with eggs, almonds, and spices to make a delicious, sweet cake. Records from the early twentieth century show that at that time the Iraqis consumed more than thirty tons of this cake each year.

Through all of antiquity, it was thought that this manna dropped from heaven. (The biblical manna of Exodus may have been the same exudate but from the tamarisk tree.) In fact, it is the product of aphids and scale insects, who insert their mouthparts into the sweet elaborated sugar and carbohydrates that run through the phloem just beneath the tree's bark. They digest what they can—primarily the scarce nitrogen—and the rest passes through them and drips from the trees.

The oak apple is not an oak fruit; it is a gall (see pages 294–301). Each gall is protected with layers of bitter tannic tissues on the outside, but the innermost chamber is full of the same sweet juices that the Kurds and the Hebrews tasted and that enterprising kids still taste today.

But both kids and Kurds have a sweeter time with the oak than did the millions who ate acorns as their staple food. Acorns smell delicious when you begin to prepare them. You crack off the shell and grind the meat into meal. It gives off a pungent, fulfilling scent, with the richness of olive oil and the tang of coffee. This fragrance is even more pronounced when you leach out the tannins with copious amounts of water and put the meal into a low oven to dry.

The taste is, to put it mildly, disappointing. One modern writer called the flavor of porridge made with acorn meal "insipid." "Absent" might be nearer to the mark. There is a definite presence and texture to the stuff, satisfying on the tongue and going down the throat. But flavor? Not a bit. Tap water is tastier.

The first prepared oak food I ate was the acorn jelly from the Korean supermarket. It was kept with the tofu in the refrigerated section, and like tofu, it came in a little square immersed in water and wrapped in a decorated plastic container. It was a lovely chocolate brown color; compared to the tofu—which has always looked to me like fat bathroom tiles—acorn jelly looked delicious.

I sliced it and ate it cold. It hit the tongue with a slimy slipperiness, like touching a slug, but fortunately, it almost immediately began to dissolve. It was lighter in texture than Jell-O, and the sensation of it, after the first shock, was pleasant. The only real trouble was that it had a flavor about as definite as the air coming out of an air-conditioner vent.

This would not do. I tried frying it in olive oil. This was better. It tasted like olive oil. I sliced it thin, grated scallions over it, and added sesame oil and rice vinegar. That was delicious, but the jelly had contributed texture and mass alone, not a molecule of flavor.

This experiment took place in the middle of the morning and was a disappointment to me. I went back to my books, wondering how anyone could have stomached stuff like this for a week, never mind several thousand years. Not that it was noxious, just vacuous. They would have died of ennui.

On the other hand, it was a great deal of fun to read about eating acorns and to talk to those who studied them. I went on with these enjoyable tasks for some time. When I noticed the time again, I was surprised to find that it was past three in the afternoon.

Never in life have I needed a clock to tell me when it was lunchtime. If I have not eaten my lunch by one or so, I begin to grow short-tempered and cranky. Fernando, the foreman at my tree company, judiciously suggests that the men may be getting hungry, in order to induce me to go buy food before I start getting angry.

This day, however, I had had a banana and coffee for breakfast, followed by six slices of acorn jelly—around four ounces—at ten-thirty. Then, nothing.

At three o'clock, I was still not hungry. Nor was I hungry at four. I went straight through to dinner without any wish for lunch or snacks, and I did not become cranky. Sometime in the middle of the afternoon, it occurred to me that the reason might be the acorn jelly. I now recognized the last sensation produced by eating it: After swallowing even the first piece, I had immediately had the pleasant feeling in the pit of my stomach of being full.

More experiments were called for. I pulled out a cookbook called *Acorns and Eat 'Em*, which had graciously been sent me by the author, Suellen Ocean. She lives near Willits, California, in the northern coastal mountains, where good acorns are plentiful. She loves frugality, and she loves eating off the land. People in Willits call her the Acorn Lady, but she longs to be known as the Betty Crocker of the Woodland. She is the sort of person who, if you called her an unregenerate hippie, might proudly nod assent.

Suellen's spiral-bound cookbook contains several dozen recipes using acorns to make everything from breakfast cereals to dinner entrées, including acorn lasagne and acorn enchiladas. There is nothing at all exotic about her recipes, no effort to imitate Native American cooking. She simply wants to make this abundant, free, healthful food a regular part of her—and of everyone's—diet.

The following Saturday morning, I made Suellen's acorn pancakes, using the acorn flour I'd bought at the Korean supermarket. Her recipe is a standard flapjack recipe, with a third of a cup of acorn meal thrown in. They cooked up a little higher, and they browned better than ordinary pancakes. Indeed, they were also a pale brown color inside. They tasted fine, a little chewier perhaps. The acorns added no flavor, but they did add that odd feeling of being very quickly filled and satisfied.

Five hours later, I still felt that way.

Next, I decided to try a recipe where acorns were the principal ingredient, not an addition to an otherwise serviceable dish. The ingredients for Suellen's acorn spinach burgers include a box of frozen chopped spinach, a cup and a half of acorn flour, two eggs, and a half cup of flour. Acorns and spinach are the main event. You make patties and cook them up in vegetable oil.

The results were not appealing to the eye. My first two burgers looked like mud pies studded with gravel, and the texture was unpromising as well. Just as I was extracting them from the pan, my wife happened in.

"Not very appetizing, are they?" I asked.

"Oh, I don't know. Let's try them," she ventured.

We broke off two shards from burger number one. There was hardly any flavor, as usual, but the texture was much more appealing than its appearance would suggest. With a little salt, the pieces were definitely palatable. Dipped in Thai chili sauce, they were tasty. Slathered with prepared horseradish, they were positively delicious.

I felt that I had discovered the first principle of acorn cookery: Mix it with something that has flavor.

I recalled reading that the Kurds still relish a dish made with acorn meal and buttermilk, and since I had some buttermilk, I added it to the acorn-spinach mix. The results fried up into tasty fritters, enriched by the pungent buttermilk. More horseradish didn't hurt either.

By the time I'd finished this experiment, I was full to bursting, though I'd eaten not much more than half a cup of acorn flour. I felt that I could eat acorn indefinitely, so long as I varied the flavorings. And perhaps too, I reasoned, since I could count on the fingers of one hand the occasions in my life when I had gone hungry, I might not value as highly the pleasant sensation of fullness as do those who have been hungry more often.

It occurred to me that acorn might well have been the foundation of all the stews and hot pots that are still the mainstays of cuisines throughout the temperate world. If indeed acorn was the staple food, it would have called out to be flavored, spiced, varied, and embellished. And if, as a number of anthropologists now think, the culminating state of hunter-gatherer culture was one in which everything from seeds and nuts to fleshy fruits, meats, fish, shellfish, turtles, insects, and berries were consumed, it would have been natural to develop a cookery based not on roasted or boiled meats but on mixed stews.

If the evidence of those who still eat acorns is taken into account, this seems to be exactly what happened. The Chinese still prepare a stew of leached acorns and water chestnuts in brown sauce. The Turks, too, still prepare racahout, a hot drink or porridge of acorn meal mixed with vanilla, sugar, and other flours. Into the nineteenth century, the Ojibway, the Menomenee, and the Iroquois of the American Northeast and upper Midwest ate acorns flavored with maple syrup, blackberries, meats, and bear oil. They sometimes mixed acorn flour with maize flour to make bread. The Apache made a pemmican of acorn flour, venison, and fat. Cahuilla cooks in southern California vary the flavor of acorn by mixing it with chia seeds, berries, or meats, then cooking the mash until it forms jellylike cakes that can be sliced and eaten exactly like my Korean acorn jelly (only tastier). Elsewhere in California, different tribes had their own favorite additions—roots, seeds, berries, and fungi. Even where they made a plain acorn mush by boiling the meal in water, they might lend it flavor by leaching the meal through aromatic leaves or by dripping the water into the leaching pit over a branch of incense cedar.

Delfina Martinez, a Pomo woman living near Hopland, California, told me about the feast days of her childhood. Her mother would build a fire and let it burn down to coals. Then she would layer wet leaves on top, followed by a layer of acorn mush, followed by a layer of salmon, followed by a layer of leaves, followed by a layer of acorn mush, and then a layer of venison. They'd keep adding layers until the pile was more than four feet high, then cover it with earth and wait until morning. All the children vied to be the first ones up in the morning, the first to open the dish and begin to eat it.

Acorn was also eaten more or less plain, as a nourishing waybread. Such breads are recorded around the world, and some are still regularly made. Sometimes the meal is mixed with clay or with hardwood ashes—a process supposed to sweeten any remaining tannins—but whatever the admixture, it comes out dark and crusty on the outside, with a spongy texture inside. The Tartars of the Crimea were still eating it at the end of the nineteenth century. In Corsica and Sardinia and in North Africa it is sometimes still eaten today. The California Indians all ate acorn breads, and the naturalist John Muir called the acorn cakes he learned to make from the Indians "the most compact and strength-giving food." It was easily portable, very nutritious, and it kept for months without spoiling.

Where acorns have been most favored as food, there have often been species reputed to be quite sweet, or at least to have definite flavor. The Cahuilla ate four different varieties, and it is said that a good cook could vary the flavor of her mush by combining the four in different proportions. Acorns from the sweetest species have evidently been eaten roasted and salted, as snack foods. In Japan and Korea, the acorn of the species called *kunugi*, a variety of *Quercus acutissima,* are sometimes so eaten, as are the acorns of *Q. emoryi* in northwestern Mexico. (The onetime importance of *kunugi* in Japan may be shown by stories that claim in ancient times there were *kunugi* so large that at morning and evening their shadows stretched for hundreds of miles.) The fruit of *Quercus ilex*—the circum-Mediterranean evergreen oak—are still a popular festival food in southern Spain and in Morocco, Tunisia, and Algeria. Up until the early years of the twentieth century, the favorite snack of the ladies at the grand opera in Madrid were roasted salted acorns.

Acorns have even been boiled to extract their oils. The oil is sometimes used as an unguent and disinfectant, and it is effective at stopping bleeding. But in parts of the globe as far separated as Morocco, Minnesota, and Mendocino, the oil has been used in cooking. In Cádiz, Spain, it is still possible to get acorn oil as an alternative to olive oil, and in Extremadura, also in Spain, there is an acorn liqueur.

That acorns are still eaten in so many ways and in so many parts of the world argues that they might once have been a staple food.

After the Ice

The later Pliocene folded land, the Pleistocene scoured it. In the Pliocene, continental movements brought tectonic plates into contact, pushing up new mountain ranges around the world. These folded into parallel ridges, much as the bedcovers make a series of ridges when you push them down from around your neck. Often, the mountains descended in cascades, from high ranges to steppes to high valleys to lower ranges down to a low alluvial valley, stepping down like grandstand seats.

In the Pleistocene, the ice flowed southward around the world, reaching its maximum penetration toward the equator by about twenty thousand years ago. Large areas of North America, Europe, and Asia were covered, though California was not and neither were Japan, most of China, and much of Southwest Asia. Average temperatures, even in unglaciated areas, declined precipitously, and the oak retreated into protected refuges. Areas that are now deserts—the Sonoran of southwest North America or the Negev in the southeastern Levant—were then covered with oaks. Major populations of oaks survived on the south coasts of the Black and Caspian Seas, and in scattered savannahs on the coast of the Levant, the northern Mediterranean, and Turkey. In China, open steppes and conifer forests dominated, while the oaks remained in the subtropical south and east. In Japan, oaks remained mainly in southeast coastal areas.

When the ice retreated, the oaks flowed out of their refuges. Frequently, they chose pathways that led along the uplands created by the new Pliocene mountains and high valleys. Out of the Negev, they cloaked the coastal hills of the Levant, traveled along mountain slopes in Syria, Lebanon, Israel, and Jordan. To the north, they met other oaks emerging from the Black and Caspian Sea refuges, and journeying along the hill country of the Zagros Mountains, in what is now southern

Turkey, and eastward into the land above the Tigris and Euphrates Rivers in Mesopotamia. Oaks rimmed the Fertile Crescent. The oaks from the Levant also turned west and reached into southern Greece. All of this happened between twelve and eight thousand years ago.

About the same time, oaks came north from the Sonoran refuge in southwestern North America and spread north along coastal and interior uplands along the length of the Pacific Coast. In eastern North America, they came north from the southeastern plains along the old lines of the Appalachian uplands, reaching into northern New England by around eight thousand years ago.

The glaciers stayed long in Europe, so it wasn't until about ten thousand years ago that the European oaks, pasted up against the Mediterranean at the edges of the Italian and Spanish peninsulas, started north. They reached Britain and southern Scandinavia within two millennia, and shortly thereafter, at the warmest moment of the current era, they even extended northward into interior Scandinavia.

In China and Japan, the oaks were the quickest to expand their range. By nine and a half millennia ago, the oaks of China had covered the central and northern uplands and reached along the coast into Korea and Siberia. In Japan, by twelve thousand years ago, the oak had reached once more through all but the northernmost islands. By eight thousand years ago, they had colonized even Hokkaido.

Oaks occupied all zones as the climate warmed, but as the temperate weather stabilized and dried, areas of the plains became less suitable to the them. Refuges like the Sonora and the Negev became deserts, and oak disappeared from them. North America's southeast piedmont retained some oaks, but they declined in density. Everywhere, however, they thrived in the uplands of the folded mountain belts.

This is the scaffold, I believe, on which human beings emerged. Not hominids, but humanity: the kind of humanity described by Pausanias, with houses and clothing and a supply of staple food. Postglacial man appeared among these stair-step mountains and valleys, ranging up the ladder to the mountains and high steppes to hunt game, such as wild sheep and goats, down to the watered plains to seek crabs, mussels, snails, and migratory birds, but living in the uplands where acorns were the staff of life, and where small wild grains fed the first domesticated animals.

Systems like this, where many different climate belts lie in close proximity—stacked one upon the other, really—and where opportunities for hunting and gathering are most various and most plentiful, are called vertical economies. Such vertical economies appeared virtually simultaneously in places thousands of miles apart: the Zagros Mountains and the Tigris and Euphrates Valley in Mesopotamia, the Jordan Rift Valley and the Judaean Hills in the Levant, the Arcadian hills of Greece, the central highlands of China, the Indian vale of Kashmir, the coast range of western North America, the highlands of Tehuacán in Mexico, the river-fen-upland complexes of central and northern Europe. Everywhere the economies were diverse— depending on a far wider variety of food sources than we do today—but everywhere they were centered on the upland belts of oak trees.

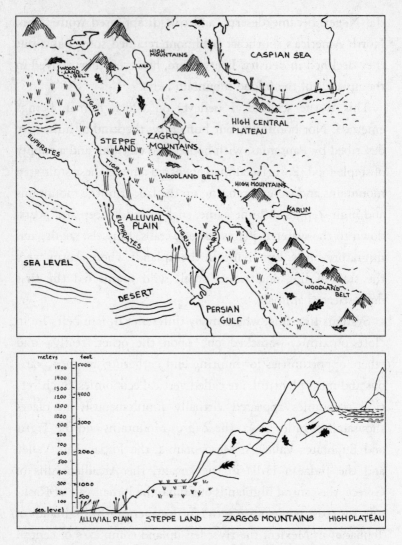

An archetypal example of a vertical economy, from plains to steppes to uplands and mountains (Nora H. Logan, after Kent V. Flannery, in "The Ecology of Early Food Production in Mesopotamia," *Science* 147:3663 [1965]: 1249)

Take, for example, the place of Eden, as it is described in Genesis. "A river rises in Eden to water the garden; beyond there it divides and becomes four branches. The name of the first is the Pishon; it is the one that winds through the whole land of Havilah, where there is gold. The gold of that land is excellent; bdellium and lapis lazuli are also there. The name of the second river is the Gihon; it is the one that winds all through the land of Cush. The name of the third river is the Tigris; it is the one that flows east of Ashur. The fourth river is the Euphrates."

This is a fairly precise description of a place with a vertical economy comprised of the Zagros Mountains down through the oak-pistachio uplands to the Assyrian steppe and finally into alluvial Mesopotamian bottomland. The settled villages that developed here were among the first in the world.

It was once thought certain that the peoples of this region at first depended on hunting large ice-age mammals. Then, when these were exhausted, they were supposed to have turned to wild sheep and goats. Later, they were to have domesticated these same sheep and goats, to make them more readily available for consumption. And finally, they were to have taken advantage of the wheat and barley that grew wild and in profusion both among the upland forests and out on the open alluvial plains. First, they were to have collected the wild grains and processed them; later, they were to have begun to cultivate them. This story was presented as a progress from the uncertain and famine-dogged life of the hunter-gatherer to the relative security of agriculture and domestic animals.

The story turns out to be almost wholly false. At the least, it needs to be stood on its head, and one major omission—the acorn—needs to be rectified. Among the largest early settlements in the Fertile Crescent was a town of thirty-two acres called Catal Huyuk, adjacent to the upland oak belt in the Konya Plain of present-day Turkey. It flourished around eight thousand years ago, just as the practice of agriculture began in the area. Archaeologists at the site uncovered both grinding implements and cement-lined subterranean storage pits. These were taken to be evidence of early farming based on the indigenous wheats and barleys. Indeed, in the later strata, not only wild strains but also genetically altered cultivated grains were found.

But there was a problem. Among the early remains there were very few if any sickles, and none of the few that were found had the characteristic sheen found on blades used to cut grass stalks. The people might have plucked up the grass plants by the root, as is still done in some parts of the world, but to do so would have meant to break many seed heads and lose the precious grain. They would only have pulled the whole plant if their intent was to use the whole plant. And the only possible use for the stalks as well as the seed heads was not as human food but as animal fodder.

If they were feeding their wheat and barley to the sheep and goats, what were the people of Catal Huyuk eating? What were they grinding with their grinders and storing in their storage pits?

They were grinding and storing acorns. The Catal Huyuk people were perhaps the last of a culture that had fed on acorns as its staple food. Only later did they reverse the feeding strategy, cutting, preserving, and threshing grains for human use, while feeding acorns to the animals. As late as the end of the nineteenth century, traveler Isabella Bishop came upon villages in Kurdistan where the people ate acorn bread, wild celery, and curds—sometimes making a kind of special pastry of the curds and acorn meal together. They did indeed grow wheat and barley, but they did not eat these. Instead, they traded the grains for cotton and tobacco.

For how many thousands of years did people in these upland belts subsist on acorns as a staple, supplemented by everything from goat meat and snails to barley tea? How secure a life was it? Weren't whole regions constantly in danger of famine? Did they not turn to agriculture for greater security?

Evidently, balanocultures were among the most stable and affluent cultures the human world has ever known. Kent Flannery, who studied the Middle Eastern peoples of this time and was among the first to debunk the idea of the big-game hunters turned to farming, wrote in a 1965 article, "Hunter-gatherer groups may get all the calories they need without even working very hard." Another researcher concluded that in the Zagros uplands near Eden it took ten times less labor to harvest acorns than it did to harvest wheat and barley. Another noted that a diet based on acorns was far more nutritious than one based on wild game, and far easier to acquire. David Bainbridge, who studied acorn use in California's surviving balanocultures—and who coined the term *balanoculture*—concluded that local oak uplands could routinely support villages of one thousand people, and these people could harvest enough acorns in three weeks to last two or three years. The nuts could be stored in above-ground aerated bins, in subterranean lined pits, or even buried at the edge of stream courses where the acorns would not only keep fresh but also be leached of their tannins by the passing water. Early European settlers in the American Northeast frequently reported that while plowing low spots near old stream courses they turned up caches full of acorns.

Acorns were nutritious, easy to acquire, and easy to store. The only laborious step in their preparation was leaching. Because of the high proportion of tannins and other astringent compounds in their meat, in order to make them palatable it was important to grind the nuts and wash them in water. Every culture where acorns were found had the technology to do this, though in southern Europe and North Africa, the acorns of the evergreen oak *Quercus ilex* were often sweet enough simply to roast and eat.

When there was plenty of oak gruel and oak bread to eat, it was unnecessary to slaughter every animal you caught. The origin of the domestication of sheep, goats, and pigs may have come because with a staple of acorns it was possible to save captured animals, breed them, and so have a "bank" of meat against possible famine. Probably, it was these animals who first were fed on the wild grains and grasses, as well as on the leaves of oaks.

With comparatively little effort, then, the people of the oak cultures could have acquired their daily bread, doing exactly as the God of Genesis specified, by eating the fruit of trees. There would have been only a basic division of labor: The men might take responsibility for expeditions in search of dietary supplements or for the more strenuous aspects of acorn harvests. Women and children could have performed the entire harvest, leaching, and food preparation. There was no shortage of oak trees in the postglacial uplands. There would have been plenty of time to stop and reflect, plenty of time to become human beings.

But if this were the case, why did anyone ever change?

Threads of Oak

For perhaps one to three thousand years, people did not change. They lived on acorns, supplemented by an increasing variety of meats, nuts, grains, spices, and herbs. It is usually suggested that human migration is caused by famine, war, disease, crop failure, or some other disaster. Perhaps in these beginnings of human cultures, it was more often caused by affluence.

Settled, peaceful society meant two things: First, it encouraged the birth of many more people. This led not only to more pressure on the food resources, but also to more diversity in the society. Different lineages developed their own aims, customs, wishes, ways of life. Because there were few disruptions, distinct groups tended to survive over many generations. Second, settled society sent the local environment into decline. Although the touch of these comparatively small groups was light compared to ours, still they required both wood for heating and for building and fodder for their animals.

The combination of woodcutting and livestock grazing could prove devastating: Cutting coppice, that is, cutting a tree down to its rootstock every ten to fifteen years, produces abundant wood for heating, cooking, and building, and given enough time, an oak will usually respond by putting out vigorous new growth and creating healthy new stems. If goats are present, however, they will gnaw down the tender new shoots again and again, until the root system runs out of energy for resprouting.

Growing population and degrading landscapes encouraged emigration. These would likely have been acts not of desperation but of boldness and vision. A daughter group would conceive of its destiny as separate from the village's and set out in pursuit of it. They would be forced neither by hostility nor by hunger. Rather, they would set out well provisioned and with great hopes. And so long as the upland chain of oak forests persisted, they would not suffer want.

The scale of this early migration is hard to exaggerate. Wherever oaks spread in the waning of the Pleistocene and the growing warmth of the Holocene, human populations could follow. The Indo-European root word for oak is *daru,* and a derivative of that word survives in the Celtic word for oak, *dru,* of which *druid* is a form. Is it possible that this word was passed from mouth to mouth and from generation to generation, from central Asia to northwest Europe, by people following the thread of the oaks?

There is no need for the impulse to have begun in one single place, once for all, and then spread around the world. It might indeed have begun in many places, wherever people first learned to leach and eat acorns. One impulse could have started near the oak refugia on the southern shores of the Black and Caspian Seas, then penetrated east along the Zagros to the Persian Gulf and south into the uplands of the Levant, eventually turning the corner into North Africa, where acorns are still common in the average diet. Another impulse could have originated in southeastern China and penetrated up through the Chinese central highlands, along the edge of the steppes and into present-day Korea and Siberia, jumping thence into Japan. A third impulse could have been transmitted from the Zagros into southern Greece and Italy, on to the Iberian Peninsula, and north into northern Europe as far as Scandinavia. A fourth impulse might have begun in the Mesoamerican highlands and moved north into California and east through the continent's southern tier, then up the Eastern Seaboard. Or was the California oak culture a product of people coming over the Bering land bridge from the comparably strong oak cultures of the Korean Peninsula?

Such transmission of human culture could have continued for millennia. The exact mix of foods would have changed according to what was available in different locations, but everywhere acorns would have been the staple. (Only in eastern North America might rival nuts have been eaten more often. There hickory nut, walnut, and butternut are tastier, as available, and as nutritious.)

Eventually, however, the spread of balanoculture would have met two obstacles. The first would have been climate. Though oaks themselves have succeeded in adapting far into the zone of snows, people are not so adaptable. More northerly balanocultures would have had to make decisions about whether to use oak as a source of foodstuff or as a source of building and heating materials. Indeed, the only places where balanocultures have survived until recent times—Japan, coastal Korea, North Africa, and California—are areas with mild climates.

The second obstacle lay behind the front of migration, in the older acorn cultures. There not only had pressure on the wood resource had much more time to build up, but the land would have continued to get poorer. Between eroded hills and voracious goats, oak woodlands began to decline. The era of plenty was drawing to a close.

Every culture has a mythic history of a former golden age of plenty. It may be that these sacred histories are, on the literal level, memories of balanoculture.

The Fall

At first, the people probably did not think that they would stop eating acorns. They were just adding other foods, either for flavor or as alternatives, maybe following a poor mast year, a season when the oaks bore few acorns. The culture of their fathers and their father's fathers had been nurtured on acorns. So would theirs be. But look: Here too was turtle soup, here were snails, here were other nuts, here was goat meat and goat milk, here was the meat and fat of acorn-fed hogs, here were grains.

One problem with balanoculture is that oaks do not bear equal annual crops. Some years are far heavier than others, and occasionally—in a season of cold and drought, say—the size of the crop might have been insufficient. Then, people might have looked to other foods: The grass that they fed to animals had edible seed heads. Parch the seed heads slightly, and the tough coverings—tasty to goats but indigestible to humans—would come off easily. The seeds could be ground into flour, using the grinders that had been invented to grind acorns into meal. The grain flour could be baked into a bread, which was lighter than acorn bread though it did not keep as well. In this way, people learned to use grains as food, even though it is unlikely that they ever would have had to depend on them.

Not until they moved away from oaks. The great flaw in balanoculture is that you can't take it with you. You have to go where the oaks are. True, in order to maintain balanocultures, customs grew up to support the propagation of oaks. In Germany and Switzerland, a law survived into the Middle Ages that required a young man contemplating matrimony to plant two young oak trees prior to the nuptials. By the time the couple's children were ready to marry, there would be two more fruit-bearing trees to help sustain them. But since the time from the planting to an oak's first fruiting is between fifteen and thirty years, the cultivation of oaks as a fruit tree has never been practicable.

The first people to move away were likely young and hopeful. Like their grandparents, who had moved from one oak grove to another in order to have their own way of life, free of the old people and their too-settled ways, this young group would have looked abroad and felt the need for a change. "Why should we stay here?" they said. "Everything is already decided here, and we have nothing to do but live and die. Let's go make our own way."

"Where can we go?" asked a doubter.

"Down there," said her friend. She would point down into the open plains, the stepped, alluvial valleys, where there were few trees but abundant grasses. "We could bring wood and build there. We would have plenty of feed for our flocks, and we could trade fodder to get acorns."

Gradually, the whole group was won over. They gathered their supplies, said their good-byes, and moved. Someone knew a good knoll with ready access to water, as well as a seasonal shepherd's hut and enclosure. It would be a good place for the village.

For the first two years, they had regular contact with their old friends up the slope. They traded fodder for acorns, and skins for timber. But it was a long uphill trek, and the old people were none too welcoming to the upstarts who'd abandoned them. Why not, said the bright woman, cook the grain right here? After all, it's all around us.

For those of us who laboriously cultivate wheat, oats, barley, rice, and other grains, it is hard to conceive of the natural abundance of the wild grains in the places where they were first plucked, threshed, hulled, ground, cooked, and eaten. Back in the 1960s, botanist Jack Harlan made an excursion to one such site in the Near East, where he found thousands of acres of uncultivated, wild emmer wheat. With a primitive flint sickle, he succeeded in gathering by hand over the course of three weeks more than a metric ton of wheat, enough to feed a large family for at least a year.

This abundance was scarcely inferior to the plenty provided by acorns. Though threshing and hulling may at first have seemed awkward and unpleasant, in the long run, they probably proved less tiring than hulling and leaching acorns. Once the change to grain had been made, why go back? The forward-thinking wishes of the leaders of the new group may have been fulfilled. Here was a new way of life, not dependent on the old ones back up among the trees.

For several generations, this new village would have prospered and grown. It might have experienced want in a bad wheat year, but the stuff was so abundant that no one would have starved. Eventually, a group of young ones would have arisen who had just the same complaint as their forebears: Everything is so settled here. We haven't got a chance. We want something different.

But when they looked out they saw neither oak grove nor hills alive with grass. They saw lowlands that were well watered with rivers and seasonal streams, but with sparser vegetation and with nothing that could serve as a staple food.

Most of them sighed. But a bright one pointed to a stream, shining in the afternoon sun, and said, "Look, we'll take grain with us. It will sprout and bear fruit in a year. Until it's ready, we will trade the produce of our herds for bread. And if we plant near the water, doubtless we will have a good harvest."

And they did. Not only did they have a good harvest, they had a bumper crop. Never had so much wheat come from so small an area. They were encouraged. They chose the best and fattest seeds to plant for the next year's crop, and again the harvest was beyond their wildest dreams. Soon, they had a village as large as the one they'd come from, gathering the wheat from one-tenth the acreage. Everyone was well fed.

But there were two unanticipated consequences. For the first time, land became valuable. The whole village's food was grown on a comparatively small parcel of land where the water table was high. (When archaeologists look at the seed caches left by such villages, they often find wheat seed mixed with sedge and other marsh grass seeds, so close did they plant wheat to the water.) And for the first time, if the water failed or flooded, the crop grew poorly or not at all, and people starved.

This was a new and not a pleasant experience. The brightest and the sweetest said, "We must see to it that there will never be a failure, and if there is, each must have at least enough to live. For this reason, I must manage the land." The most unscrupulous said, "We must see to it that if there is too little for all, we, at least, will have enough. We must possess this land."

For the first time, a small number of people acquired control over the land that produced the food for everyone. This too was likely not a pleasant experience, especially for those who were less well provided for in times of famine.

The brightest of these, driven by need, thought, "Look, we can go out onto the plains where streams run a long ways. True, it is hotter and more exposed there, but we will dig channels from the stream out into the land. This way, the water will be spread far wider, and we will be able to grow so much that everyone will have enough to eat again."

Notice the "again." This is perhaps the first time in human history that a people looked back in longing to a previous age of plenty.

But on the flats, all went well at first. A great work of digging and cultivating was established. Improved grains that had been developed in their native villages were applied to a vast area and produced immense amounts of food. Soon, they could thumb their noses at the floodplain cultivators. Where the ancestral village might have been able to support at most a couple hundred people, the irrigators could support several thousand in one town.

In the ancient Near East, this was the root of the great civilizations of the plains: the Assyrians, Babylonians, Akkadians, and Sumerians. They raised heretofore unimagined amounts of food on a very small amount of land. Perhaps 10 percent of the land supported all of the people. Towns turned to walled cities, and pyramids rose. The best land fell into the hands of a few people, and these people—for the common good or for their own good, and usually for both—arranged the lives and possibilities of the rest. A third of the food came from land belonging to only 1 percent of the people. There was little choice for the discontented, since without these food resources, they would die.

The same two consequences—ownership and starvation—quickly visited the cities of the plains. There was a third as well: salt. It turns out that when flatlands that have no natural route for drainage are irrigated, sooner or later salts concentrate in the soils. When they come near the surface, they poison wheat. Indeed, later food production in all of these early cultures was skewed toward barley, which is more salt tolerant.

Perhaps in this situation, the least well fed would have looked back in longing to the lands of their forefathers, now practically lost in legend, when they had freely eaten the fruit of the trees. But it was too late to go back.

For many generations, the oak-wood dwellers had been

degrading their own lands. When they cut oaks for firewood or to build their houses, it was hard for the trees to get a fresh start, because goats would eat the tender shoots. Furthermore, the more successful the people and their grain-eating neighbors became, the more the population grew and the more they would need oak for its wood, not its food. After all, they could trade the wood for grain and become more modern.

Now the cities of the plains made large demands on the upland dwellers for wood. They had houses, temples, and walls to build. They needed charcoal to forge first bronze, then iron, and to make pottery and glass. They had a lot of interior space to heat. Furthermore, they could enforce their demands with weapons that a culture of gatherers could not resist.

Today, in regions of the Zagros Mountains in Iran, where ten thousand years ago people gathered in a few weeks all the food they needed for the year, people eke out a hard and bare living farming wheat and barley. They labor a hundred times as hard for fewer calories than did their primitive ancestors. The hills are bare; the land eroded. The oak trees are almost all gone.

The Place That Waits for Me

Only a single constellation of balanocultures survived into historical times, the cultures of Native California. These were still intact when the first Europeans arrived in the late eighteenth century, and remnants of them survive to this day. There was nothing particularly notable or memorable about these cultures—nothing at least most people would call memorable. They had none of the fierce nobility of the Plains tribes, the clear intelligence of the Cherokee, or the civilized life of the Hopi. The first Europeans to arrive described the natives as sleek, well fed, and indolent. "Diggers" the Gold Rush immigrants called them, noting that they were adept at finding edible and medicinal roots. The immigrants found this disgusting, although it was in every case a part of the broad-spectrum gathering habits that made the natives far better adapted to life in that land than were the invaders. With few exceptions, the tribes welcomed the invading Europeans, and without exception they were overwhelmed.

Yet there is one remarkable fact about them: There were so many of them. Within the boundaries of present-day California, there were at least one hundred separate tribes, each with its own language (or at least dialect), stories, religion, customs, and territory. Each territory was a classic vertical economy, with mountain, foothills, and watered valleys, each zone contributing to the tribe's subsistence. The boundaries were usually defined by the watershed of a given river. Tribelets—often groups of a few extended families—lived in small villages by the rivers and traveled up-country for hunting and gathering. Most tribes gathered all the tribelets together at least once a year for a ceremony of world renewal, but the dances, songs, costumes, and settings of these ceremonies were widely different from tribe to tribe.

Virtually every tribe, however, had one thing in common: acorns. Every territory contained oaks in its foothill and/or mountain section, and the acorns gathered were a central feature of the tribe's livelihood. Other gathered foods varied widely—from salmon in the north to pinole nuts in the south—but acorn was everywhere a staple food. It is even thought that the presence of sufficient acorns within a given area was what allowed a tribe to settle down and become a tribe, that is, that the distribution of oak trees determined the total number of tribes. Early European travelers came to recognize how close they were to a village in the middle of any given day by the continual dull booming thump of women driving mortars into pestles to grind the acorn meat into meal.

Throughout California, the annual autumn acorn gathering was the year's main event. Often, the whole tribe would go together into the hills and establish an acorn camp near the grove that was theirs by right and custom. Among the Cahuilla of southern California, a family called its holding *Meki'i'wah,* which means "the place that waits for me."

The rituals for claiming a tree were various but everywhere polite. Among the Wintu of northern California, if you found a tree that you thought no one else had found, you could lean sticks all around it to signify your right to pick its acorns. On the other hand, if you had seen this tree before but failed to mark it, you could remove one of the other person's sticks and claim a part of the tree for yourself, but you would have to bargain with the person to pay for your claim. In any case, if a tree were particularly large, you would not likely be so greedy as to claim the whole thing. Instead, you would lean your stick against the tree to mark a single branch.

When the tribe arrived at the acorn camp, they would set out in parties to their various trees and groves. The men would climb the trees to shake off the acorns, while the women and children would gather both the freshly fallen and those that were already on the ground. (The former made the best soup, smooth and white, while the latter made a darker and less desirable soup.) To strip two trees in a day was thought to be fast work, since a large tree might contain three hundred to five hundred pounds of acorns.

Not everyone spent all day gathering. The men might hunt acorn-eating animals that were attracted to the groves for the same reason as the humans. Others would gather wood for firewood or for paddles and handles. A shaman would collect certain oak galls, which were powdered and used against eye infections and to help heal wounds. Bark was gathered to use as a fuel and to boil to extract a black dye used in basketry. Leaves were taken to be dried, shredded, and used as tinder.

Some tribes started shucking acorns right at the base of the tree, others carried the day's haul back to the acorn camp. There, the adults would sit around the fire, cracking shells and extracting the meat. Some did it with their teeth, crunching the shells first lengthwise and then crosswise; others used rocks. Other acorns had only their caps removed; they were taken intact back to the village for long-term storage. Some young women strung whole acorns into necklaces, and children played games of jacks with them or did juggling tricks.

For two or three weeks, the group stayed at the acorn camp. After the first day, at least one woman would stay in the camp to guard the acorns that had been spread out to dry. Her job was to turn them so that they dried thoroughly and to ward off predators. Although she did no collecting, she was allotted a share of the harvest.

A California native woman shucking acorns (C. H. Merriam, courtesy of the Bancroft Library, University of California, Berkeley)

It was a hard walk back to the main village, since each family had to carry hundreds of pounds of acorns, but at least it was mainly downhill. At the village, some of the acorns were stored in raised bins woven out of stems and vines and coated with pine pitch or lined with bay laurel leaves to discourage marauders. Others were buried whole in a bark-lined pit or underwater. After the destruction of the Native cultures, for decades farmers working the low ground—just as they did in the Northeast—would still plow up caches of these water-buried acorns, which were still sound and edible after half a century.

Communal mortars and pestles in a large rock
(C. H. Merriam, courtesy of the Bancroft Library,
University of California, Berkeley)

Back home, life returned to its daily round. Everyday by midday, pestles boomed in mortars as the women ground the acorns into meal. Usually, a woman had several mortars and pestles. Some mortars were hollowed out in the bedrock, and some she could carry with her. She made these by heating the stone, then chipping it away with a sharp rock. Young women did most of the grinding, while the older women sifted the meal and divided it into different kinds by size—some for soup, some for mush, some for bread. (You were not supposed to sing love songs while grinding, because the words might cause the pestles to break.) In most tribes, only a woman or her daughters had the right to use her mortars and pestles. When she died, one of her mortars would be broken and buried upside down.

After grinding the acorns, the women leached the meal, usually in a depression made in sand and lined with leaves to keep the sand from mixing with the meal. Cool or warm water was poured over the meal until the bitter tannins were washed out. The process took two hours or more. With certain acorns, especially those from the valley oak, *Quercus lobata,* the Wintu women would soak the leached acorn meal in still water until a mold began to develop. The mold was said to make valley oak acorn bread particularly tasty.

A few tribes leached whole acorns to eat whole, but almost all of them made their acorns into soup, mush, or bread. The main difference between the three was the consistency of the meal mixture, and the final product depended on a woman's skill with rocks and water. To make soup, a small amount of coarser meal went into a basket with water. Hot rocks were lowered in to heat and thicken the soup. With more and finer meal, the same process yielded a jellied mush that, if allowed to set, could be cut into squares and distributed like a flan or Jell-O. Alternatively, it could be eaten warm, and during festivals it often was; people dipped their hands into a common pot and licked them clean. Acorn bread was often a woman's finest dish, since it involved getting just the right consistency of mush, then laying it on a bed of fern above hot rocks, then covering the mush with flavorings and with more fern and finally earth. The bread would cook overnight and be ready in the morning. Depending on the flavorings that had been added, it could keep for months.

In 1907, C. H. Merriam of *National Geographic* attended a funeral that was held in a round house in Tuolumne County, California. He reported that to feed the guests, the cooks began preparing several days prior to the feast, leaching acorn in four 5-foot diameter leaching pits. When the festival began, they had ready fifty huge baskets, each containing one to two bushels of fresh acorn mush. They had also prepared fifty loaves of acorn bread. He estimated that the guests at the two-day feast consumed upward of a ton of acorn.

The amount itself is staggering, but what is more wonderful to me is that this was nothing but a late—perhaps the last—example of a tradition that extended throughout the length and breadth of California, virtually unchanged, for more than five thousand years. The great thing about balanoculture is that it allowed the slow maturation, in place, of at least one hundred separate cultures. Protected by sea and high mountains and deserts from alien encroachment and defended from themselves by a temperate climate that made it unnecessary to deplete the oaks for fuel, the California cultures proliferated.

Were they great cultures? By our standards, they were not. They left no great monuments. They could not defend themselves from the invaders who at last arrived. In a sense they were repulsively bland, as the Gold Rush prospectors found them, though to be honest, it was the prospectors themselves who should have gone by the perjorative "Diggers." Their digging destroyed whole landscapes in pursuit of gold, while the natives lived lightly and without harm on the land.

Nevertheless, even A. L. Kroeber, the great anthropologist who devoted his life to the study of the California Indians and to gathering all available information about their lives, agreed in the end with the historian Arnold Toynbee, that the Native Californians could not have had great cultures, because their wishes met with too little resistance.

I think Kroeber made a category mistake. The cultures of the California Indians were not greater or less great than other cultures. They were something different altogether.

In 1911 a lone man was caught raiding a chicken coop in northern California. He was naked, long haired, and carried stone tools. They threw an old coat over him and called the University of California. Kroeber and his wife, Theodora, appeared to take the man to live with them in Berkeley. His name was Ishi, and so far as anyone could tell, he was the last living member of his tribe, the Yahi.

Theodora Kroeber's book about Ishi's life with them is one of the classics of California literature. In it, among many other things, she recounts some of the stories and legends that Ishi told them.

One tale concerns an old woman who tells her daughter, "There is a strong, good-looking young man. I want you to go and marry him." The daughter refuses, even though the man is interested in her, until her mother threatens to throw her out of the house. "How will we live," the mother says, "unless this man comes to hunt for us? I am getting too old." The daughter consents and seeks out the man. They then play the courtship game in which she again and again invites and rebuffs him, until finally she agrees to marry. "But listen," she says to him, "you have to go and ask my mother's consent. Maybe she won't want you in our family." You can hear the laughter of the Yahi people who listened to this story around the fire for perhaps several thousand years. And you and I laugh just as much, with just the same understanding that this is the way that human relations are. As a modern girl once put it, "He came after me. I ran and I ran, until I caught him."

The mother in the story, of course, is overjoyed. She describes the life that will blossom among them as though she were weaving it on a loom: "He will come to live with us and bring us meat while we cook him acorns. Then babies will be born, and you will make your own house, and I will come to you. You will rejoice to see me coming with a basket full of acorns on my back."

It is hard to conceive of a story that better expresses the wishes and the understandings of everyday life. Before and beyond all struggle, what every human seeks is the peace of such joys as lovers, parents, children, and friends experience. Nothing beyond.

The first aliens to penetrate the California balanocultures might never have eaten an acorn. But they came in boots tanned with oak bark, eating pigs fed on acorns, carrying iron weapons forged in smithies fueled by oak charcoal, writing with ink made from oak galls, and riding across the seas in oak ships. They were restless seekers, like so many Europeans of the time, looking for a new life, new wealth, new worlds.

The newcomers and the natives shared one thing: Both flourished on account of oak.

THE AGE OF OAK

WINDSOR IS NOW a bedroom community for London. It so happens that the queen sleeps in one of the bedrooms. Windsor Castle is there. The hostess of my bed-and-breakfast had my day all planned when I came down to breakfast. As she served me sausage and eggs, toast and fried tomatoes, she told me what time the castle opened. I told her I was there to see trees.

"Trees! What trees?" she said, not sure if she'd misheard me.

"The old oaks," I said. "The Dotards."

"Oh, those!" she said. "They might be gone."

I started.

"No, no," she corrected herself. "I'm wrong. They wanted to widen the A332, but people wouldn't let them. Because of the trees."

The A332 leads out of Windsor immediately into beautiful open land, with the Great Park, the huge royal estate, on both sides. There are long vistas of lawns and trees. Fields fold gently

An ancient oak in the Windsor Great Park
(Courtesy of English Nature)

into the horizon. I saw some respectable oaks far off. I hoped those weren't the Dotards. They weren't.

The ancient oaks were right beside the road, but I did not recognize them at first. They didn't look like trees. They looked like old slag heaps with a smoke of dark green leaves. They stretched away for hundreds of yards down both sides of the road. They must have once been an immense allée, each tree ninety feet tall with crowns almost as many feet wide, covering acres with their broad shade. Now, the tallest could not be more than thirty feet. Some were dark hulks with no leaves at all.

How beautiful they were, these ancient trees. Up close, they were not monolithic at all. They were stag-headed and wholly alive. More than anything else, their trunks looked like flames fixed in an instant. Great curving hunks of blond deadwood rose up out of the dark furrowed bark and flowed on upward around gaping holes and fissures. The broken trunks were massive, each more than six feet in diameter. Somehow, from the tops and sides of these ruins, great branches sprouted. Some of the branches must have weighed more than three tons.

Here was a huge bare knuckle of wood twelve feet high with just two branches left to it. I felt how smooth the bare wood was, and how rough the bark that remained. I plucked gigantic fingers of splintered wood and made deep twangs—thong, thong, thong—that echoed through the hollow boles.

Here was a tree whose trunk seemed to have steps. Between the gnarled burls and the places where wounds had healed, you could walk up into the tree as though up a ladder into an attic full of forgotten treasures.

I did.

There were still half a dozen large branches on this tree. When I climbed up among them, I realized that I was completely concealed from people out in the field. It was nearly noon on Saturday. People ate their Marmite on crackers and sipped their tea. I sat invisible in my broken, ancient oak.

I remembered reading in the legends of oak that the Scots patriot William Wallace had concealed himself with one hundred of his men in a single oak tree. Inconceivable, I had thought. But from this vantage point, it was clear that there was no exaggeration. An open-grown, ninety-foot-tall oak, with its vast, S-curving branches reaching sixty feet away from the trunk, its leaves thrown to the outside where they could catch the sun, would indeed be capable of hiding scores. I felt sure that even this ruined trunk could have concealed two dozen.

But how could these old trees still be here at all? They were probably part of the first oak plantation in England, sown in the Great Park by Lord Burleigh in 1588. Shakespeare was a young man when they were planted. It had been decades since they even looked like trees, but the English had foregone widening a road in order to protect them.

People admire oaks, even dying ones. There is no more common plant in heraldry, where oak signifies strength, good character, and stability. The strong and faithful in Britain are said to have "hearts of oak." Oak was Zeus' tree to the Greeks, Jove's tree to the Romans, Thor's tree to the Norse, the Dagda's tree to the Celts, and El's tree to the Hebrews. Isaiah, prophesying the redemption of Israel, wrote, "They will be called oaks of righteousness, the planting of the lord, to display his glory" (Isaiah 61:3).

In any landscape, oaks are the keepers. They make a mark against the sky, so that even without knowing it, people find their way by them. They are part of the gestalt by which the eye guides us. Because they are so large and steady, oaks are also easy to find. The hollows in them have been used as post office boxes around the world, and people have hidden precious items in them. Surveyors use them as landmarks and line trees, because they expect the trees to last.

Old oaks acquire such fame that when they fall, people go into mourning. The Charter Oak near Hartford, Connecticut, where colonists hid their charter from the English redcoats in 1687, was officially mourned when it failed in 1856, and it is still the state tree. Maryland's Wye Oak was purchased by the state in 1939 and made the center of a park in its honor. It was 96 feet tall and probably more than 460 years old when it failed in a storm in 2002, but not before it had been perpetuated by cloning. At Penshurst Place, the estate of the Sidney

A whole town seems to live beneath the branches of the Cowthorpe Oak, then the largest surviving oak in England, in this nineteenth-century engraving
(Collection of William Bryant Logan)

family in English Kent, stands a blasted oak trunk with one living branch, said to be the tree to which the poet Philip Sidney fled for peace and quiet back in the 1570s.

Generosity and hospitality are attributed to oaks. In medieval Irish law, oak was listed as the first of the seven chieftan trees, for its size, its beauty, and the generosity of its fruits. (Stern penalties were invoked for the unauthorized felling of oaks.) Hesiod praised the generosity of oak, for he claimed that it yielded three fruits—acorns, honey, and galls—where other trees yielded but one. Philemon, a poor man, once sheltered the gods Zeus and Mercury, who were traveling in disguise. In thanks for his hospitality, the gods changed him into an oak when he died.

Brute size was never a claim made for oaks. Henry David Thoreau, indeed, praised the little shrub oak, a scrubby, stiff-branched, spreading oak of the New England woods, calling it "rigid as iron, clean as the atmosphere, hardy as virtue, innocent and sweet as a maiden." He began one journal entry with the observation, "I should not be ashamed to have a shrub oak for my coat-of-arms."

The strong and proud might even be brought to grief by oaks. Greek and Roman storytellers delighted to tell about Milo of Croton, the most powerful man of his day. An invincible warrior, he could kill an ox with a single blow and eat it on the spot. One afternoon, walking in the Sila woods near Croton, in the heel of Italy's boot, he came upon an oak tree that the woodcutters had left for the day, with wedges still in its trunk. He decided to finish the job himself. But when he put his hands into the cleft and started to pull, the wedges fell out.

The oak snapped shut on his hands, cuffing him to the tree, where later that night the wolves found him. Lord Byron adapted the story in the early nineteenth century, giving Napoleon Bonaparte the role of Milo and Britain that of the steadfast oak.

Admiration of oaks goes as far back into the human past as there are records. A medieval Welsh poem, "The Battle of the Trees," here translated by Robert Graves, is thought to collect ancient traditions about the symbolic character of plants. In it, the poet envisions the trees coming to battle:

> *With foot-beat of the swift oak*
> *Heaven and earth rung;*
> *'Stout guardian of the Door,'*
> *His name in every tongue.*

A great deal is packed into these four short lines. The oak is swift: It appears when it is needed. It makes itself felt in heaven and earth. (Virgil records the mistaken belief that oak was the only tree whose roots were as deep underground as its branches were high overhead.) The last two lines notice something strange: In almost every language with Indo-European roots, the name for a door is derived from the name of oak, *duir*— door, tur, puerta, porte are all examples. Granted, oak is stout; granted too, doors have very often been made out of oak. But why "guardian of the door"?

Among the ancient Celts, the months were named for trees. *Duir* was the name of the month that led up to the summer solstice; *Tinne,* or holly, was the name of the month that followed solstice. Though this tradition in England referred to the true holly, *Ilex,* in most of early Europe, where the deciduous and the evergreen oaks existed side by side, it referred to the ever-green *Quercus coccifera,* the Kermes oak, whose leaves exactly resemble those of English holly. (Both also had red "fruits," the ilex having true fruits and the oak often bearing red galls, which were the source of red dye for royal robes.) The two months, then, were originally oak months, the first of the deciduous and the second of the evergreen oak.

Oak guarded the door of the year, when the waxing part of the year ended and the waning part began. No other tree known to the Celts had both evergreen and deciduous species, never mind one so large, so various, and so generous. The deciduous oaks watched over the waxing year, and the ever-green oaks over the waning year. Together, they were a guaran-tee that life could pass through winter and be renewed. In older legends, the oak is identified with Janus, the two-faced Roman god of the portal. Looking both ways, the oak knotted past and future into the ongoing present.

Many rituals seem to have acted out this sense. Sir James Frazer's *The Golden Bough*—the first and most influential book about ancient ritual in the European tradition—claimed to reconstruct a number of them, including the ritual murder of the "oak king" of one year by the oak king of the next year. (The book's title refers to the oak bough decked with yellow-green mistletoe that Frazer supposes to have been cut from deciduous oaks as part of such a ritual.) At least one such rit-ual survives in literary form as the medieval story of Sir Gawain and the Green Knight.

A giant clad top to bottom in green appears at King Arthur's court during a feast on the shortest day of the year. He is holding a club covered in holly leaves, and he has a large ax. He offers to let any of King Arthur's knights cut off his immense head with a single blow of the ax, providing that knight will give the Green Knight the right to do the same to him one year hence. Sir Gawain takes up the challenge.

With one blow, as agreed, he neatly severs the Green Knight's head from his body, only to find that the giant is unfazed. Picking up his head in his arms, the knight declares that he will see Sir Gawain again in one year's time and rides away.

After a year of quest, during which he shows both his knightly qualities and his weaknesses, Gawain faithfully appears before the Green Knight and bares his neck. The giant takes a mighty swing with the ax, but pulls up at the last moment, only grazing Gawain's neck, causing the blood to flow. At that, Gawain starts up. One blow was the bargain, no more. But the giant doesn't seek a second. Instead, he praises Gawain for courage and honesty and lets him go free.

It is very likely that this story represents the survival of a ritual involving the evergreen oak, the Green Knight, and the deciduous oak, Sir Gawain. One stays green all year despite the weather—it cannot be killed—but the other is saved from death and recovers, although in winter it had seemed quite dead. Gawain's courage, generosity, devotion, and charity are what save him, in short his knightly virtues. To the people who fully understood this story, the virtues of oak pointed out what was virtuous for men.

It was not a sentimental connection. Until the last century, people understood the lives of trees, because people depended upon trees. They saw how persistent a young oak was, how it would sprout back again and again, even if an animal ate the entire shoot. They saw how a forest of oaks showed many differences in leaf, in bud, and in acorn among the individual trees, as though each had a name and a character of its own. They noticed that oaks put out leaves several times per year, not just once or twice. They observed that the oak seemed to be constantly full of birds and animals and insects and worms, yet it did not die of them. Indeed, the rest of a forest's trees seemed to draw back and make room for an older oak, like courtiers stepping back and bowing at the entrance of the king. These people used every part of the oak, from wood to the leaves to the bark to the acorns to the galls.

People knew what oaks were like, and they wanted to be like them too. Among the Celts, the Druids served as lawgivers and singers. The honorific name *Druid* is derived from two words: *dru,* meaning "oak," and *wid,* "to see or to know." The people who were thought fit to sing the tribe's stories and to interpret its laws were said to "know the oak."

Oak Knowledge

Where is the cradle of Western civilization? Most people would say it is in the Mediterranean. But if we are talking about the turbulent, world-spanning civilization of which we are almost all now a part, then its cradle was not only in the blue, dolphin-flecked waters of the Mediterranean, but also in the gray, choppy waters of the North, where the boats go out for whales, for herring, and for cod. Here, the oak's lessons of cooperation, flexibility, persistence, community, and generosity were well learned—of necessity in such comparatively unforgiving lands—and oak itself became the prime material of the crafts that helped hominids become humans.

I stood at the very end of the Oslo Fjord looking out into the North Sea. The day was very cold, and last year's sere yellow-brown grass and copper-colored heather sighed and hissed in the gusts of wind. It was February in southern Norway, and even at midday anything standing cast a long dark shadow on the ground. We had passed the last trees some few miles back, about the place where I'd noticed a sign that read MOOSE CROSSING. No oaks here. Just the end of everything.

There were no headlands to speak of, no dramatic cliffs. The land just petered out in water. To the right, I could see the jagged inlet of a deep cove; beyond that were two or three more, not the same shape but the same idea. To the left, the deep fjord stretched across to Sweden, but the land beyond was much too far away to see. The whitecaps looked as if they ought to be made out of snow or ice.

I came over a low rise and saw the last beach. Not sand. It was composed of pale, rounded rocks, the largest the size of bowling balls and the smallest as big as golf balls. But in the distance I could see something quite strange. Someone had piled mounds all along the beach. The biggest were about ten feet high and seventy-five feet long. The smallest were a foot high and a few feet long, joined together in a chain that stretched away down the beach. Some were strict piles, others were outlines with hollow centers.

All were in the shape of boats.

I stepped off the grassland into the stones and almost fell over. This beach was not just a layer of rocks; the rounded stones went deep into the earth. I tried to dig down to see, but my gloved hands numbed before I had reached more than a foot below the surface. Nothing but sea-smoothed stone.

A funny place for a cradle of civilization, but such it is. "What the temple was to the Greeks, the ship was to the Vikings," wrote Johannes Bronsted in *The Vikings*. These stone boats were created in the early Bronze Age, more than four thousand years ago, and they could be seen by people miles offshore. Within five hundred miles of the spot where I stood— along the tortuous shores of the Oslo Fjord, the North Sea, the Baltic Sea, the Irish Sea, and the English Channel—was born the impulse that would later carry men for the first time around the world, knitting the globe into the single entity we naturally perceive it to be today.

Bronsted might have added not only the boat, but home, and not only the Norsemen, but all the peoples whose lives gave onto these gray seas. In a place of lonely land fingers and fen-bound islands, home was indeed a haven for those coming in off the sea. In the north, even the queen wove for her family and poured out the mead at the feast. It was not beneath the lady of the house to serve. The English language today is more than half Latinate, but the old English and old Norse *house* and *home* are the words we use. Only lawyers use the Latinate *domicile*. If you ever hear that word, you are probably in a fix.

The climate of the emerging Holocene made the boat and home important, and oak became the prime material of both. The melting of the ice sheets, which began about fifteen thousand years ago, set off many processes at once. Sea levels rose as waters that had been trapped as ice now flowed into the ocean. Yet for thousands of years, it was still possible to go dryshod from France to England, from England to Ireland, from Asia to Canada.

The oaks did, and so did people. About ten thousand years ago, the first colonizers returned to Britain, which was not yet an island. They were Stone Age peoples and deciduous oaks, the ancestors of *Quercus robur*. They both followed the coasts, where the climate was mildest and the variety of food—shellfish, fish, game, acorns, berries—was widest.

Memory of this fast-changing landscape is preserved in stories. In the tale of Branwen, Daughter of Llyr, in the *Mabinogion*—probably first set down in writing around the eleventh century A.D. but referring to events long before than time— King Bendigeidfran (King Bran in other Celtic stories) leads an expedition from Wales to Ireland, across the Irish Sea, which is then but a pair of rivers:

In those days, the deep water was not wide. He went by wading. There were but two rivers, the Lli and Archan were they called, but thereafter the deep water grew wider, when the deep overflowed the kingdoms.

The ice kept melting, and the sea kept rising; coastlands gradually drew back before the rising water. Watersheds were drowned; mountains became islands; valleys became fjords; coastal plains became shallows. Forests were drowned. Britain became an island, then Ireland. Nine-thousand-year-old oaks are presently being recovered from a depth of 120 feet in the Irish Channel, huge trees that do not even start to branch until ninety feet up the trunk. In drowned settlements on the Baltic coast are found flint spear points among the blackened oak trunks. One oak recovered from a Pembrokeshire bog had a trunk eleven and a half feet in diameter. The whole margin of Europe today contains what the British call "Noah's woods," the remnants of ancient forests and the source of bog oak.

Not only the soils but the seas were young. And shallow. This was to be a land of few natural deep-water harbors. Increasingly, the rising water cut off the way through valleys and made high ground into coastal jetties. Every family living on their land spit needed a boat to get around on or a bridge to cross. They needed a homestead, sturdy, modest, durable, and above all independent. The boats had to be light and strong enough to beach. The bridges had to withstand wet and dry for many years.

The Holocene affected the other cradle of civilization, the Mediterranean, too. In the last quarter of the twentieth century, marine geologists discovered that about 5600 B.C., the rising waters of the Mediterranean breached the narrow land dam at the Bosporus, near present-day Istanbul, plunging into a huge freshwater lake that quickly became the saltwater Black Sea. (The ancient freshwater is still trapped at the bottom of the sea.) The Flood is not mythic, but historical. It happened.

When the Bosporus was broken, the whole ancient, coastal culture of the Black Sea—favored by warm climate and regular, abundant rains—was drowned. For two years the roaring waters poured over the breach at a rate of ten cubic miles of water per day. Each day, the water drowned another mile of once-fertile territory. The freshwater ecosystem perished. The old coastline was drowned 450 feet deep beneath the new sea. The peoples of the coast fled, losing almost everything they possessed. As the story is remembered in the Mesopotamian *Epic of Gilgamesh*—among the oldest works of literature extant—the gods warned Utnaphishtim, "Give up possessions, seek thou life! Forswear your goods and keep the soul alive!"

Some believe that this event set off the migrations that created the Mesopotamian and biblical civilizations and that sent the Celtic people west and then north into Europe.

North or south, as the Holocene matured, humans went where oaks were. Among the oldest stories preserved by the Greeks is that of Deucalion. Just after the great Flood, it is said, he ascended to the oak forest at Dodona, and there in thanksgiving for his safe escape established the oracle of Jupiter that would last until the fall of Rome.

It is told that, ever after, the oaks of that grove had the power of speech; the oracle interpreted the sound of the leaves. When Argo built a boat for Jason to sail into the new Black Sea in search of the Golden Fleece, he is said to have used the oaks of Dodona for the mast and the beams. The ship's timbers would speak to warn the Argonauts whenever danger was near. It was said that this was the first boat ever made that was plank-built and not simply hollowed out of a log.

The Holocene is the age of oaks and humans. Never before had there been such a thing as memory or culture, or such a thing as a hearth, or such a thing as stock-raising or house-building or shipbuilding. Hominids had been around for at least two million years, but they became human beings as they began to learn to use oak.

Humans arose through the exercise of memory, reason, and skill in a world that was warming and blooming. Memory gave to reason its material, and skill proved a thought to be true (or not). Skill suggested a turn of the wrist, memory preserved it, reason compared it. The human attributes—courage, faithfulness, patience, fortitude, prudence, honor—were forged where the faculties met, in the making of home in the world.

With and through the use of the resistant materials of the earth—oak prominent among them—humans exercised their faculties, showed their qualities, and spread and multiplied until they covered the whole planet.

From as early as humankind has left remains, there is evidence of the intimate relationship between people and oaks. This was true not only in the old Mediterranean landscapes, protected by their bowl of a sea, but especially for the peoples of the west and the north, who were exposed both to harder climates and to more unforgiving seas.

In "Reginsmol," a verse tale from the tenth-century *Poetic Edda*, comes the following dialogue that represents the strength of coastal storms in northern waters:

> *"Who yonder rides on Raevil's steeds,*
> *O'er towering waves and waters wild?*
> *The sail-horses all with sweat are dripping,*
> *Nor can the sea-steeds the gale withstand."*

"On the sea-trees here are Sigurth and I,
The storm drives us on to our death;
The waves crash down on the forward deck,
And the roller-steeds sink. Who seeks our names?"

All those steeds and sea-trees are the oak-built longships of Sigurth and Regin; equally remarkable is the fact that in this howling gale, men are represented as having a conversation. As the tale goes on, the men ride out the storm, thanks to their wooden ship.

The northern peoples built their civilization out of oak. Even a polished stone ax could take down a moderate-sized oak, and seasoned oak wedges then split it. As long as the oak logs were fresh cut—what foresters call "green"—they were also easily split into lengths along the lines of the prominent annual rings and along the radial rays to make plank-paved boardwalks and sturdy boats. Oak logs could be squared and framed, and joints cut in the strong wood held buildings together. Humans ate food made from acorns and fed them to their pigs. Indeed, pigs were domesticated from wild animals that roamed the woods feeding on acorns.

Human beings learned from the woods around them. The more people used oak, the more they found it could do. The wildwood of postglacial Europe had more lindens, more elms, and more beeches than it had oaks, but as humans made use of the wildwood, the percentage of oak in the forests increased. Other woods were good for firewood, poles, fences, and wattle walls, but oak was valued also for the bigger tasks: roadways, frames, doors, palisades, henges, barrels, coffins, boats, tanning, ink. The material necessities for settled human life can all be made from an oak tree.

More than six thousand years ago, men were already foresters. They had by then invented the system of coppice and standards, a practice that gave them the wood they needed without destroying the forests.

Standards were the trees that were meant for load-bearing structural timbers, whether for houses, boats, bridges, wagon beds, barrels, or axles. Oak was almost always the preferred standard tree. They would be cut only once every fifty or one hundred years. Other trees (and some oaks, of course) would be cut to the ground, or *coppiced,* every five to twenty-five years. The fresh sprouts were grown to the desired size, then cut again. In this way, humans could get the daily wood—firewood, fencing, poles, and stakes—that they needed, as well as bark for tanning. At the same time, the woods could be preserved and renewed.

In wood pastures, where cattle, sheep, or goats were run in among the trees, the farmers learned instead to pollard the trees, cutting them to about shoulder level instead of to the ground. The new sprouts were protected from the eager mouths of the animals.

Though oaks were often left to grow large—sometimes for a century or more—they too could be coppiced and pollarded. Oak firewood was among the best. A cord of oak wood—a stack four feet wide, four feet high, and eight feet long—releases twenty-three million BTUs when burned. Even in a minimally efficient hearth, this represents the equivalent of more than one hundred gallons of number-two heating oil.

Any wood could be tossed into the fire, of course. None but oak was allowed to start the need-fire. Throughout the length and breadth of Europe, the ritual fires at midsummer and in midwinter, and whenever disease threatened the flocks, were kindled by the "need-fire." The customs differed from place to place—in one place two naked men had to start it, in another two young boys—but always it involved twirling an oak pole pressed into a hole drilled in an oak plank until the rubbed wood smoldered and tinder lit.

This "forced fire," *teine-eigen*, was the first fire of each new year and the one from which all other fires in the community were to be lit. A girl who jumped back and forth three times across the fire would be fertile, a sick cow who walked through the ashes would be cured. The first boy to bring a flaming brand of it back to his own house brought luck to the house for the whole year. Ash from the fire would ward off lightning.

Enclosure was the first principle of settled human dwellings, not only of the house but of the household, including out-buildings, wells, farmyards, and woodpiles. Oak made fences and walls that kept stock and children in and held enemies out. Splints of cleft oak were driven into the ground and rods made of hazel or other flexible species from the understory of the oak woods were woven among the stakes to make a fence that the pigs could not breach. Whole or split oak poles, driven into the ground close together and bound with withies made palings and circular monuments called "henges."

Fencing was second only to fuel as the most common use of wood. In France during the Middle Ages, the punishment for the most heinous crimes might be simply that the householder had to leap over the homestead's fence and never return. If the fence were too high, he would have to pole vault.

Oak made travel through bogs, swamps, and fens secure. From late Neolithic times, men learned to cut oak planks and joint them so they held together securely. The first boardwalk bridges were made of these planks, either laid end to end and side to side or notched and fit into supports, to make a more stable surface.

The first buildings and furniture were made of oak, using the joint that came to be called mortise and tenon. The mortise consisted of a hole carefully chiseled either part or all the way through one member. The second member, the tenon, was the male piece, cut as exactly as possible to fit into the mortise. In this way timbers could be joined into four-square frames, effecting a solid combination that could stand up and bear weight.

Planks made of split oak trunks transformed the world. Tubs and barrels were made of cleft oak. So were looms. In the Far North, they began to make plank-sewn boats by joining long, cleft sections of oak and waterproofing the joints with pitch. Eventually, the Vikings perfected this craft, using the oak's suppleness to form their longships, boats that wriggled like eels through the waves, their timbers bending with the shifting stresses but never breaking.

Even oak galls were useful. One was boiled to yield a pigment used for painting and for dyeing clothes, and later to make the finest drawing and writing ink. Oak bark and a black oak gall were soaked to get the acids to tan leather and so make shoes, reins, and saddles. Tanned leather also made vellum, the north's first paper.

For ten thousand years oak was the prime resource of what was to become the Western world. Through *Dru-Wid,* "oak knowledge," humans learned to make homes and roads, ships and shoes, settles and bedsteads, harnesses and reins, wagons and plows, pants and tunics, swords and ink. Far from being a place of fear or darkness, the oak wood was the sustainer of people's lives.

The Sweet Track

If you lived in western Europe about six thousand years ago, you could not do better than to live near a fen, a spring-fed coastal marsh. It had everything a person could want. The surrounding uplands or fen islands had oak and lime and ash and alder, for shelter and for fire and for acorns. If you cleared the wood, you could make open fields to feed cattle and attract game. The fen edges dried out in summer, making good summer pasture, and the reeds could be cut to thatch the roof of your house. Best of all, in the depths of winter, when for others there might be little or nothing to eat, you had eels and fish, ducks and geese that you trapped in the waters of the fen. Salt could be pan-dried from the brackish water nearest the sea, and it could be used or exchanged with foreigners for remarkable objects, such as polished axes. People were not likely to go hungry on a fen.

Indeed, the fen was the coastal equivalent of the vertical economies of the Mediterranean and the Near East. Even when agriculture came to western Europe, it was more an addition to the existing patterns of broad-spectrum gathering and hunting than it was a change of life. Cereals were grown in the uplands, but agricultural practices focused on livestock: cattle, sheep, and pigs. These fit neatly into the existing gathering culture, so long as there were not too many animals. Pigs ate the acorns, sheep and cattle grazed the cleared fields, all left their dung to preserve the soil's fertility.

But there was one serious drawback to living on a fen. If you wanted to go anywhere, you had to go around. There were no villages, to speak of, so far as we know. The inhabitants occupied individual farmsteads, usually on a well-drained gravel island or upland edge. To visit your neighbors could be an arduous undertaking, even if they lived within waving distance.

One day in the spring of 3807 B.C., a group of neighbors— probably not more than a couple dozen adults, all told— decided to do something about it. They had acquired good axes—not the old flaked stone axes of the Mesolithic and Paleolithic, which could be ruined with a few unlucky strokes. A flaked flint ax is full of little ridges where the flakes came off, and each of these ridges is a point of weakness along which the ax can chip or even break in two. People in the Somerset Levels, located on the coast of what is now southwestern England, were having none of such nasty, untrustworthy tools. They acquired polished stone axes, axes that after the flaking had been laboriously ground and polished against sandy substrates until their completely uniform cutting surfaces shone.

So beautiful and useful were these tools that they became the most desirable possession a person could have. Their polish was kept up over years and they were traded over vast distances. One ax, probably intentionally dropped into a Somerset Level fen as an offering, had been made of jadeite and brought from the foothills of the central European Alps. Later in the Neolithic, such axes were principal cult objects, and some were even worn as badges or necklaces.

The reason for this may not be far to seek, if we look at those families on the Somerset Levels in the year 3807 B.C. Taking axes in hand, they decided to bridge the shallow water that stretched for almost a mile between their houses. To do this, they started by felling alder, lindens, and young oaks. They dragged these to the fen and started to lay two parallel tracks with these slender logs, the first as the base of their new road-way, the second the rough track for the workmen to use. They laid the logs straight and true, from one dry place to the other.

Then they cut stakes, again of solid round young trunks, mainly alder. Because of the uniform size of the stakes, it seems probable that these people were already using a coppice system, intentionally cutting hardwoods to the base, waiting seven, ten, or twenty years, and then cutting down the new trees that had arisen. They placed these long stakes in an X; the bottoms were driven into the peaty earth, the bottom of the crossing of the X rested on the straight line of poles, giving the structure stability.

To finish the road, however, they did something that had possibly never been done before. They used axes to open splits in the ends of large oak logs. Then they used wooden mallets and seasoned oak wedges to split out narrower pieces from the oak halves and quarters. They could do this because of oak's prominent rays. Rays, lines of cells that run at right angles to the wood's growth rings, provide planes that, if you are careful and you learn to read the grain of the wood, can keep the splits straight and true. The cleavers kept on with this splitting until they no longer had half logs, quarter logs, or any other kind of log. They had planks, some of them up to fifteen feet long and three feet wide.

These planks, resting in hollows of the X's, made a flat surface upon which the families could securely walk. When a piece seemed a little unstable, the workers learned that they could cut a notch into the edge of the plank, axing across the grain and then wedging out the broken fibers, and fit the notch into a stake driven straight into the marsh bottom or nest it against the X supports.

You and I live in a world of things that are flat and oblong. The page you are reading is such, so is the book when you close it. Your front door is such, and so are the windows. The framing of the house, whether it is wood or steel studs, is the same, and so are the chimney bricks. The folded sheets in your closet are flat and oblong, not to mention the tablecloths, handkerchiefs, and basically your shirts and pants and skirts and blouses and overcoats. The desktop is so, and so are the shingles and the floor tiles and the floorboards. The road is such and so is the bridge rail.

In 3807 B.C., until those people finished their track, the only things that were straight and oblong were the bodies of water—pretty, but if you tried to cut a piece of it, you got nowhere—and certain rocks or landforms, also pretty, but equally hard to move though for opposite reasons. (Not that rock was not tried, as the barrows and Stonehenge show.) Nothing else looked at all like it looks today. These people didn't go for a walk in the woods to escape the presence of man-made objects. They were bringing the first man-made objects out of the natural world.

Our world was born of their acts. A man named Sweet discovered this trackway in a peat bog in 1973, so it is called the Sweet Track. The anonymous men and their families who made it spent about a year cutting and moving wood. They built the whole thing in a day.

The Sweet Track was in use for more than a decade. But it was succeeded by miles of trackways built throughout the lowlands there in Somerset and throughout northern Europe, for three thousand years. These were the first bridges or highways.

In the years that people used the track, they threw things from it. Perhaps these were offerings, perhaps accidents. Regardless, they tell us what people did and what they valued. A pot full of hazelnuts proves that farmers were still gatherers as well. Was she taking the hazelnuts to her friends across the track? Was she offering them to thank the Lady of the Lake for the track and her hazels and her friends?

A box carved of oak is another artifact, and from the look of it, it was carved precisely to fit an ax whose haft was found not far from the box. Was this left by the worker accidentally? Was it offered? How important to people was an ax if they kept it safe in its own carved box?

A third artifact is a child's toy, also cut out of oak. It is not absolutely certain, but it is very likely that the toy was also an ax.

Rings and Rays

When prehistoric Britain is mentioned, the first thing that people think of is Stonehenge. Maybe they see it full of white-robed Druids practicing obscure sacrifices, or they may see it as a primitive observatory meant to mark the seasons and the stars. What people seldom realize is that Stonehenge is by no means unique. It is one among a large number of ring monuments, collectively called henges, at least four dozen of which have been found in Britain. An aerial survey suggests that there may be many more. All were built in the late Neolithic and early Bronze Age, between 2500 and 2000 B.C.

Most of the henges that have been excavated were largely or entirely made with oak posts, stood on end and fixed in the ground. Those popularized in the press have picturesque names, like Woodhenge or the Sanctuary or Seahenge, but many have precise archaeologist site-names, like Durrington Walls IV or Site IV Mount Pleasant. Most are composed of concentric rings—sometimes six or more—that once contained tall oak posts. Because they are on dry land, not in fens or bogs, the posts have long since rotted away, leaving only debris in postholes. It is not known whether these henges were roofed, though the fact that in some of them the posts obviously leaned outward before collapsing suggests that the weight of a roof might have encouraged the collapse.

At the entrances to these monuments and at particular places within them have been found artifacts pertaining to almost every aspect of Neolithic life: antler picks, human bones, flint scrapers and flakes, axes and balls carved in chalk, the bones of pigs and cattle, arrowheads, and quantities of pot fragments. The pots were almost all of a type called Grooved Ware, which leads archaeologist Richard Bradley in *The Passage of Arms* to suggest that the henges were made by a particular people among all those then living in Britain.

Apparently, the people who built the henges brought all of the products of their lives—foods, craft objects, hunting tools, digging tools, cutting tools, and even the bones of the dead—into the henge monuments. Yet the greatest artifacts of all are the henges themselves.

The henges were complex in conception and in execution, far more so than the Sweet Track. They were complex not only to build but also to use. The concentric rings were often offset one to another, so that it was hard to walk in a straight line from the outside to the center of the structure. Even in the henges that have cruciform aisles, straight paths into the center are often blocked by an extra post or a stone. The idea, evidently, was that a person had to wander inside of them. To go in or out, you also had to go around and around.

Henges are found only in Britain, but they are similar to other shapes that begin to appear in the iconography of Europe and the Mediterranean at about the same time. In Brittany, the rock art began to show not only representations of things in the world but also strange concentric patterns that are like those seen today by persons in trance states. These shapes and patterns are called "entoptics" because it is thought that they are created within the eye itself. (They are something like the circles that you see when you stare at a bright light, then close your eyes.) In the Mediterranean, the pattern of the labyrinth—reniform drawing, that is, kidney-shaped with winding interior paths—is scrawled and painted. Here too is a shape that cannot be directly penetrated but must be explored through many turnings.

Henge monument design bears a strong resemblance to one thing in nature: the end of an oak log. The pattern of annual rings and radial rays suggests a mixture of the straight way and the curved. And working with oak would have reinforced this knowledge of complexity. You could split it into halves or quarters, but you could also take slices out of it, by means of the rays, or cut it tangentially along the annual rings. So much variation came from this one solid object that its pattern of organization may well have impressed itself on the people who observed it.

The human mind grew through the knowledge born of craft. With the felling of oaks, the making of axes, the cutting of planks, the planing of boards—not to mention the making of pots, the domestication of cattle, the technologies of the bow and the spear, the clearing of forests and the creation of coppice—a change had taken place in the human relationship to nature. Hominids were no longer simply extracting subsistence from the natural world, they were transforming what they took into stable, ordered forms.

In order to make, they had to imagine. In order to imagine, they had to make. Between thinking and making, the world we know emerged. The mind made making possible, but making opened the mind. Who knows if the imagination of the plank came before the making of the first one, or if the form of the henge was first conceived or observed? To imagine and to make are reinforcing pairs.

The great thing about the henge, the entoptic circles, the labyrinth, is that you turn and wander through them. They are context, an inside, a place where you are alone with yourself. They are monuments about the mind.

Some think that these shapes hint at esoteric quests, but it seems clear to me that they celebrate the growth of the human mind, that is, the growth in the human understanding and use of the mind. They acknowledge the genuinely human conviction that there *is* an interior world of thought and reflection that is woven into the exterior world of events and weather and life and death. They represent a milestone in the humanization of hominids, a process that began long ago but is still ongoing.

The lately found Seahenge is odd and not technically a henge, but it strongly makes this point. Located on what was once a high spot in a salt marsh near the sea in Norfolk, the monument consists of a single circle of closely spaced, split oak logs. The oaks rotted and fell long ago—only pieces remain. They were about eight feet tall. Until 1998, the monument was preserved under the shallow waters of the North Sea.

The half-logs were set with their bark side facing outward, their cut sides facing in. There would have been no chance of squeezing between the logs, and entry was only through an opening formed by a fork in one of the logs. It has seemed to some that Seahenge is simply the remains of a round house, but little household debris has been found there. Instead, at the very center of the circle is a very large oak stump, set upside down so that the root flare faces upward, radiating in an ordered but tortuous pattern.

It is tempting to suppose that the monument represented a shorthand for the more elaborate henges. The multiple rays and rings were suggested by the oak trunk itself, and the mixed pattern of movement was suggested by the flaring roots. It would pose the same questions as the more formal henge: What is our inner world, and where are we bound?

Four Million Pieces of Wood

The largest fen in Britain is located near Peterborough in Cambridgeshire. Four thousand years ago, the fen occupied about one million acres. Fed by both flood tides and two rivers, as well as springs, it was a rich landscape, especially for livestock farmers. On the dry land, wood could be harvested and grain could be grown. Meadows were good for cutting hay. Flats that dependably flooded in some seasons but not in others made fine, self-renewing pastureland. From the wetlands came reeds for thatch, as well as snails and all kinds of minor foods; occasionally in a dry year, these lands might be grazed. Out on the water were fish, eels, and a huge variety of wildfowl. The inhabitants ate, at least, goose, mallard, cormorant, pelican, heron, stork, mute swan, barnacle goose, teal, duck, merganser, sea eagle, goshawk, buzzard, crane, coot, and crow.

But first of all they ate steaks, chops, and stew, because they raised thousands of head of livestock, mainly cattle and sheep. There is evidence of a system of paddocks and pens fit thickly around the fen edges, which brought Francis Pryor to study the site in 1971. As he and his crew located boundary ditches and excavated to Bronze Age soil levels, they exposed the most elaborate system of ancient livestock culture yet found. The fields and paddocks were shaped and organized as precisely and to the same design as today's livestock pens.

Flocks would be driven from the seasonal fields to dry, enclosed pasture. There were collecting pens where the animals would mill about, complete with a side gate to permit sorting out of the males. Females were then driven into a funnel now called "the crush" and then into a race, where they had to pass single file. The gradual narrowing was designed to produce the calming effect that tight grouping has on herd animals. In addition, the long race allowed animals to be examined one by one. At the far end of the race was a triple gate system that allowed the stockmen to separate out the young, the breeders, and the culls.

The system was not intended for just a few animals. It was scaled to accept herds of up to two thousand. Provisions for watering were evidently not meant to support large herds over the long term. This then was not a small family site, but a place where groups of farmers could meet to exchange animals, to talk, and to feast. It was probably in community stock pens like these that the northern European (and American) predilection for eating large quantities of meat was born. It was a sign of wealth to be able to consume big steaks and chops. In many quarters, it remains so today.

One late afternoon in 1982, at the tail end of a cold foggy November, Pryor was walking along the top of a dyke with a few of his colleagues when he tripped over the end of a piece of wood that had been excavated from a lower level in the course of ordinary drainage work. He cursed it for yet another bit of leftover medieval sluice-gate material and yanked the piece from the ground. He observed that it was tangentially split oak. Intrigued, he slid down to the base of the ditch, where he tripped again over a slender protruding post. Pulling it from the ground, he observed that it too was oak, which surprised him because it felt much too light to be oak.

Looking up, he estimated how deep in time he was by how deep in the ditch he stood. He was clearly well below the Romano-British level, meaning that this wood might well be Bronze Age oak.

Cold and wet as it was on the fen in late November, he felt that the matter deserved further investigation there and then. They dug around and quickly began to come on more oak, including a long plank into which a clean square hole had been cut. The hole was penetrated by the end of another piece of oak, which had been cut to fit the hole. This find excited them, because it meant that whatever they had found, it was at least partly in its original place, and it was an object that had been *built*.

What was it? The question percolated in Pryor's mind. He thought that it might be a Bronze Age trackway across the fen, a great-granddaughter of the Sweet Track a portion of the hundreds of miles of trackway known to have criss-crossed Bronze Age Britain. Even so, it was exciting to think that they had discovered a path that led somewhere. By tracing it, they could perhaps learn more about how people had lived on this land.

The first task was to find the track's edges. That would tell them the track's direction. A few days later, in spite of the wet, frigid weather, Pryor returned to the site with six colleagues and six shovels. They started to move mud away from the post they had found previously, expecting to come on the track edge within a few feet to either side. But the track did not seem to have an edge.

Pryor sent out colleagues to scout the area. Sixty feet to his right, one of them came on another oak plank. Ninety feet in the opposite direction, another ran into more oak boards. All the diggers reported finding wood at the expected depth. This was an unexpected result, however, because no one had ever heard of a trackway that was so wide in so many different directions at once.

More scouts went out. More reports came back. Wood. Wood. Wood. More wood.

In fact, they continued digging for most of the next decade before they finally came up with a size for the thing they'd found. It was an artificial island in what was then the middle of the fen, about five and a half acres in area. The island was made wholly of wood, most of the structural planks oak and most of the roundwood underpinning alder.

But questions remained. As Geoffrey Wainright, then chief rescue archaeology inspector for English Heritage, the major funding source for such digs, put it, "It seems to be very important, but what is it?"

And the island was by no means the only thing. There were tracks leading to it from both sides of the natural bay. On either side were the dense stock pens that Pryor had come to study, and indeed, the tracks connected directly with at least one of the main drove ways.

But adjacent to these tracks were also up to five lines of upright posts driven into the fen bottom. There were at least two thousand, ranging in size from a few inches to more than a foot in diameter. They had been placed upright in the fen in roughly five different lines. The first line—built mainly of alder wood—had been set probably around 1250 B.C. The others had been placed in succeeding centuries during the more than four hundred years that the site was used and repaired.

One or more of the lines were comparatively widely spaced. The masses of equal-length roundwood pieces found mixed behind these lines suggest that the posts were connected by wattle-and-daub panels. This was the stuff that most walls would be made of in England for the next three thousand years, a latticework of small wood, branches and later split oak lathes, overlaid with mud plaster. The effect would have been to wall off and protect a bay of the fen from the open water beyond.

The backbone of the structure, built about 1000 B.C., was a tightly packed line of oak trunks. This wall was only broken in one or two locations along the entire length, and on its bay side ran a path of oak planks set atop a roundwood base. A less tightly packed line of trunks flanked the track on the other side. Around the same time, a rough line of irregular trunks angling out into the open fen was placed beyond the principal track, and beyond that irregular line, another regular line of posts was joined with wattle-and-daub walls.

All in all the wooden island and the tracks and posts of Flag Fen comprise about four million structural pieces of wood. The Eiffel Tower, by contrast, has about eighteen thousand structural members. Whatever it was, this Bronze Age monument was important and elaborate.

There is plenty of evidence that people feasted there, including the discarded bones of sheep, cattle, and pigs, but none of the daily household refuse that would suggest anyone lived on the site. On the other hand, a large number of objects were thrown into the fen, probably from the island, the tracks, and the surrounding shore. Often, the objects were broken before being thrown, or only a part of a complex object was dropped.

Virtually all of these objects are items of craft: a group of grinder mortars called *saddle querns* (none have their pestle stones), bent bronze swords and a pair of bronze shears, bent awls, brooches and pins, tin wheels and earrings, a broken fleshhook, ruined razors, gouges, a bronze ax with its haft sheared off, a broken necklace of polished shale, spearheads, daggers, broken jars and cups. Why?

Certainly, they were offerings. Archaeologists often note the most famous example of throwing a sword into the water—the return of King Arthur's sword Excalibur to the Lady of the Lake—but they also can point to an unbroken tradition leading from the Neolithic to the Bronze Age and the Iron Age and beyond of throwing precious objects into fens, lakes, pools, wells, and even rivers. Thousands and thousands of objects. Indeed, the wooden structures at Flag Fen were evidently used for this purpose for more than four hundred years, until the waters finally rose high enough to drown the tracks and the island. Even then, people continued throwing things in from the fen edge and from the riverbanks for another thousand years.

Almost every one of us is a distant participant in this long tradition. Who has not thrown a coin into a well or a pool, making a wish over the offering, watching the coin as it fell to see if it spun and where it settled? But there are many more links to the ancient past that have survived even into the twenty-first century. There were until modern times more than three thousand "holy" wells, springs, pools, and bogs in Ireland alone. Clothing, pins, jewelry, buttons, buckles, brooches, coins, and little stones were regularly thrown into them or piled or hung beside them.

Many of the stories of pools and wells also include a ritual during which livestock are washed in the water or at least sprinkled with it. Up until the early years of the twentieth century, for example, men and women still celebrated each year at Padstow Traitor Pool by running a hobbyhorse through the town's streets and then taking it to the pool for a drink. After it had drunk and a song had been sung in its honor, the people would be sprinkled with mud and water from the pool, and the evening would, as one observer put it, "end in riot and dissipation." Such rituals may well go straight back to the Flag Fen culture of more than three thousand years ago and further back into the Neolithic.

But the principal offering at Flag Fen—the main object of craft left in the water—was the monument itself. It is possible indeed that part of the post lines or the platform itself might have been laid in succession as offerings. Studies of tree rings on wood at the site show that all the wood was cut nearby, and all evidently worked and prepared in order to be used at Flag Fen.

The timber structures at Flag Fen offered to the water the

best oak craft that the people could show. Though it is not certain how the welter of wood was assembled in the days of its use, there is nevertheless plenty of evidence that the makers there had become carpenters. They had invented the basic joints that allow any building or piece of furniture or cart or ship to stand up against gravity and storm.

Structure is second nature to us. Indeed, we go to the wilderness to escape from the constant sight of things that are designed and built. But building was new in the Bronze Age. It might often have been a matter of simply pounding wood into the earth or laying planks atop a layer of sticks, but here it was also a matter of securely joining one piece of wood to another.

With bronze axes, it was much easier to cut precise mortises and tenons than it had been with stone. The oak would still often be split radially using wood wedges, or carefully split tangentially with the same wedges to get a wider plank, but the crosscuts would be made with a sharp bronze ax, the waste wood removed with bronze gouges. The archaeologists' illustrated catalog of the recovered wood is an anthology of mortises—round, oblong, slotted, and squared—with their corresponding tenons.

Here is one of the first records of lap joints, scarf joints, and housing joints. These are joints that in any woodwork allow pieces to be joined at angles and end-to-end, so that a structure can be longer than the longest single board available, and so that a roof might be raised and a structure braced diagonally to prevent its collapsing in wind and storm.

Lying in piles, these long-fallen oak planks hardly look

prepossessing—nothing like a pyramid or a Parthenon—but they are of comparable importance. Here, in oak more than three thousand years old, are the joints to make the house, the bridge, the bed, the chair, the boat. For though many of these are not even made of wood anymore, never mind oak, they use joinery principles first created by working with oak. Even the great Sarsen stones of Stonehenge were fit together using carpenters' joints.

Bargeroosterveld

A Neolithic person sees or imagines a plank taken from the trunk of an oak. Out of nature he makes a plane. Bronze Age people explore the cylinder by making upstanding henges out of oak, and the rectangle by making tracks and platforms. But to use these posts and beams of oak to make a cubic frame—an open solid not found in nature—takes another step in imagination.

At Bargeroosterveld, near Emmen, in the province of Drenthe, the Netherlands, in 1957, peat diggers came on evidence of such a revolution. It was tiny. The whole structure fit inside a stone circle about thirteen feet in diameter. But it was a frame made of oak, built about 1475 B.C., and it contained every carpenter's joint needed for houses or anything else that would be made of oak for the next three thousand years. Not only that, it had been beautiful.

It had a foundation of oak planks laid in a rectangular pattern on the surface of the bog. There had been posts at the cor-

Bargeroosterveld as it looked when it was whole
(Courtesy of Tjalling Waterbolk)

ners, each seven to eight feet high. The builders had squared
the edges of these posts, so they looked like any 4 x 4 that you
might buy at the lumberyard, but they had tapered the bottoms, so the posts fit into slots cut in the foundation. Four
inner posts, not squared but rounded, were also fitted into the
foundation.

Four beams joined the tops of the all the posts into an open
rectangular frame. Each beam was squared, but the ends of
each had been carved into light, rising arcs, so the structure

must have appeared to have horns or wings. At each corner, three pieces of oak had to meet: one post and two beams. All three had to be cut and fit together so that they acted as a unit. Gravity must not be allowed to pull them down, nor wind part them.

What's the big deal? you may wonder. It is nothing but a simple wooden frame. All you have to do is shape narrower ends on the posts and cut holes in the foundation and in the beams so that they fit over the posts. Of course, if you want to make it last, you probably also need to cut matching notches

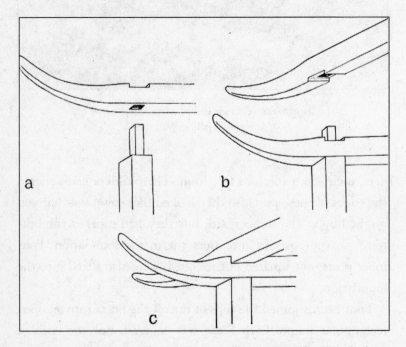

The joints that hold the frame together (Courtesy of Tjaling Waterbolk)

halfway through each of the beams, so that they marry where they are joined atop each post. Otherwise, a knock or a gust of wind might make a parallelogram of your rectangle. But it is a simple structure nonetheless.

Something similar was said to the architect Filippo Brunelleschi about two millennia later, when he refused to submit his design for Florence's duomo to the building committee. He finally agreed that he would show them the design, provided they could stand an egg on end on a table.

Each member of the committee tried it without success while the architect looked on. Confessing defeat, the committeemen demanded that Brunelleschi perform the feat. He promptly knocked the bottom off the shell, and stood the rest of the egg on "end."

"Oh that!" they exclaimed. "Anybody could have done that!"

"Yes," replied the architect. "But none of you did. It is the same with my design. Anyone could have thought of it, but I did."

You may say that this little Bronze Age pergola is far from the magnificent duomo, but they are similar in principle. It took a shaping imagination to envision each and then to find out how each could be realized in space. Oak, because it could reliably be cut and shaped in so many ways, made the frame found at Bargeroosterveld possible.

What was the structure for?

The foundation planks were oriented roughly north and south. Between them, on the east and west sides, the building had been closed by four twelve-inch-diameter logs and splits of oak. These were disposed in a long rectangle just inside the load-bearing posts, leading the archaeologists to speculate that they had originally been placed to be the legs of a table whose surface has long since decayed. It is possible, they speculated, that the building was used for a funeral, a body exposed on the table until the birds had reduced it to clean white bones.

But they found no evidence of human or animal bones on the site. Nor did they find any "ritual deposition," any objects of craft that had been offered at the bog. All they found were a great number of wood chips, indicating that the oak had been shaped and finished on-site. Furthermore, the way in which each of the winged beam ends had been broken strongly suggested that the structure had been intentionally wrecked.

Perhaps it was not made for use at all, but as an offering. The Egyptian pyramids at Giza were already more than a millennium old when this modest building was erected (and destroyed). The great pyramid is about 481 feet tall, where the thing at Bargeroosterveld is just above a tall man's height. The pyramid celebrated power. The elect, the intelligent, and the god-gifted directed their servants to push, pull, scrape, and cut. The building at Bargeroosterveld celebrates the dawning of the power in the hands of the individual to make beauty. No slaves were needed to erect it. Free men could and did follow this tradition of building from that day to this.

But it appears that the makers also destroyed their creation at Bargeroosterveld. Among humans, the long tradition of thank offerings often includes mutilation of what is offered, partly as a guarantee that the object will not be taken up and used again. In just this way, the objects in Flag Fen were bent or broken. One thinks also of the sand mandalas, the elaborate paintings made by Tibetan monks using colored sand, that are erased shortly after completion.

To make a frame out of the resistant materials of the world is a defining human activity. The name of the gods to whom such actions were dedicated is almost immaterial and in any case forgotten. But one can see why human memory suggests that oak was worshiped: It was the first durable material in the West that could be transformed by individuals into shapes that they saw in their minds.

The Dehesa

Forests seem permanent and natural. Few are. For at least six thousand years, men and women have shaped the woods. The forests as we know them are products of craft.

The people who built the Sweet Track already were cutting coppice, a principal forest management practice that survived down to our time. To make coppice is to cut down a broadleaf tree—oak, ash, thorn, hazel, alder, or elm—and wait for it to sprout three, four, five, or more new shoots. Coppice shoots grow more than twice as fast as the original wood, so in seven, ten, twelve, or even twenty-five years, the cutter would have very nice poles or beams. These he would cut, harvesting the wood for firewood, charcoal, fence posts, palings, or the foundation course of the Sweet Track. The stool, the base of a coppice tree, would sprout again, indefinitely. Coppicing may actually prolong the trees' lives. A typical ash, for example, will live two hundred years uncoppiced, but five hundred to one thousand years under coppice.

Some say that the people of the Sweet Track did not intend to make coppice, but merely took advantage of the sprouts that grew when they cut down large trees to open land for crops and pasture. However this may be, it is certain that their descendants on the Somerset Levels, in about 2500 B.C., were using a sophisticated coppice system. When they built the nearby Walton Track, the people made it out of hurdles woven from slender hazel poles. All the poles are of more or less the same diameter, whether they were three years old or seven years old when cut. They were selected out of a wide area of hazel coppice for their size alone, a sophisticated procedure called "drawing." Prior to this discovery, no one thought that drawing had been practiced until the early Middle Ages.

It may well be that Walton Track builders were only partly thinking about the size needed to weave neat and lovely hurdles. It is likely that they were also cutting the young wood in order to feed the leaves and tender twigs to their stock. In early forest management, nothing went to waste.

Up until our times, a forest was not a wilderness. From it might be harvested timber, firewood, tanbark, charcoal, fence and hurdle poles, splints for wattle, grass to pasture sheep or cattle, acorns to fatten hogs, honey from beehives, ink from oak galls. In return, the users left the forests the dung of the grazing animals. An ecologist would look on it as an exchange of organic carbon.

The woodspeople developed a fine, supple intelligence, born of their intimate knowledge of the trees and wood. They knew, for example, that almost no conifer would resprout once it was cut to the ground, but that yew was an exception. Furthermore, they observed that coppiced yew stems were tough and flexible, and made a sturdy cord. The first boats not hollowed from logs were made with oak planks sewn together with yew thread.

The rise of iron, pottery, and glassmaking industries is often blamed for the destruction of European forests. In his *Ancient Woodland*, Oliver Rackham estimated that the six main ironworks in the Weald of Roman Britain made about ninety thousand tons of iron between A.D. 120 and 240, or about 550 tons per year. Eighty-four tons of wood made enough charcoal to produce one ton of iron; this means 7.5 million tons of wood were needed in a little more than a century. Had they cut only new trees, they would have had to destroy seventy-five thousand acres, or 9 percent, of the region's forests. Using a coppice system, however, they could have returned again and again to the same woods, and this is indeed what was done.

The real destroyers of the woods were the plow-using farmers, who spread through Europe, beginning in the Iron Age. A gardener using a hoe or a digging stick works around the tree stumps, allowing them to regenerate, but the plow farmer wants to make long, straight rows. This farmer didn't just cut down trees, he grubbed out the stumps. It is estimated that half the woodland in lowland Britain was destroyed in the seven hundred years between the coming of the Iron Age and Roman occupation.

In the Domesday Book, compiled for William the Conqueror in A.D. 1086 to estimate the wealth and set the metes and bounds of his England, there is a plow tally: eighty-one thousand. Early in the medieval period, largely thanks to these tools, only 15 percent of the land was forested, and it was rare to find a wood that could not be walked across in an hour.

Nevertheless, the mixed-use forest held on as long as did the commons, the lands owned jointly by the people of a village or town. But when the commons were seized for large-scale agriculture—a process that culminated in Britain with the enclosures of the eighteenth and nineteenth centuries—many of the remaining forests were seized and removed. Large landowners who held forests of their own came to regard them as "excrescences." Wheat brought a faster and higher cash return.

Scientific foresters were not much help. They reasoned that oak was a drag on profits, because it grew so slowly. They counseled their clients to convert their forests to fast-growing conifers, which produced sawtimber better and faster than oak. "Fir buys the horse, oak buys the saddle," was the saying.

But woodland management did not disappear everywhere. In a few places, it evolved uninterrupted over a period of more than six thousand years. Some few of the *dehesas* of southwestern Spain have been maintained since the time of the Sweet Track.

A good *dehesa* looks like a park. The principal trees are evergreen oak, *Quercus ilex,* or, where the soils are sandier, cork oak, *Quercus suber,* or a mixture of both. Under the trees are grasses. Paleontologists have estimated the antiquity of the *dehesa* system by testing fossil pollens. If they find pollen from oak and shrubs, the normal components of an oak woodland in that part of the world, they figure that there was a relatively natural ecosystem on the site. When they find oak and grass pollen, they believe they have found a *dehesa*. The evidence for the oldest *dehesa* dates to about 4100 B.C.

Dehesas efficiently produce more per acre than any system of modern agriculture, but the products consist of small quantities of many things, not a huge amount of one thing. In spring, cattle or sheep graze the grasses. It is important not to overgraze, so managers keep a running idea of how many cattle the land can support: if the soil is thin, perhaps one cow per ten to twenty acres; if the soil is better, maybe one per four or five acres.

Trees are pruned, usually on a seven-year cycle, to maintain the open structure that the managers believe results in better fruit production. The prunings are used for firewood or to make charcoal. The charcoal from the *dehesas* is highly valued in the towns and cities. Where cork oaks are present, cork is harvested on a nine-year cycle, by stripping the thick outer bark from the tree. A cork harvest looks like mountains of warped boards, but it is transformed into bottle corks, insulators, gaskets, cups, and plates. The bark of *Quercus ilex* may be harvested as tanbark.

Since the Neolithic Age, crops have been planted among the trees. In the oldest *dehesa*, grape pollen was more abundant than oak, suggesting that wine grapes were intensively cultivated there. More commonly today, chickpeas, broad beans, wheat, or lupine are grown in the sunny openings.

The average *dehesa* yields about six hundred pounds of acorns per acre each year. A few of these still go to make local specialties such as acorn liqueur, sweet acorns for snacking, and a candy made of acorn-stuffed dates. But pigs get most of the mast. From October to January, Iberian hogs are turned loose in the *dehesa*. The average hog doubles in weight during this time, gorging on fallen acorns. The hog leaves droppings in exchange. From these hogs come Serrano's cured hams.

In short, a single tract of land provides grain, beans, cork, charcoal, tanbark, and the meat and skins of goats, pigs, and sheep. In some *dehesas,* there is also a small amount of kermes oak *(Quercus coccifera),* the galls of which are used to make inks and dyes.

No one gets rich on a *dehesa*. It isn't managed for high cash output, but it does sustain both the forest and the people who use it. It is not easy work. Don Quixote, whom Cervantes made a native of this region, waxed eloquent about the good old days of easy pastoralism, when people and pigs alike fed on acorns. But that was fiction.

The *dehesa* required of Neolithic man—and of everyone thereafter on that land—a complex code of use allied to half a dozen crafts. The very word *dehesa* is derived from *adehesamiento,* the name for the thirteenth-century town land grants that gave these domesticated woodlands official status in the medieval world. Both those words are derived from *defensa,* meaning "defended." The term specifically set apart *dehesas* from other land and demanded a system of care and protection.

On some *dehesas,* that system lasted well into the last quarter of the twentieth century. Often, the *dehesa* still belonged to the town or village in common. The rights to do different things there—to pasture cattle, to prune, to take cork, to fatten hogs, to grow crops—might be auctioned off periodically at the town hall, or in a few cases, these were rights taken in turn by the local citizens. Not all the pruning or all the cork or tanbark would be scheduled for any given year, so the wealth was spread around. This meant that one's fellow citizens were always on the watch to see that rights were not being abused. Everyone suffered if the land declined.

Thus, most families in a town needed to keep up skills as farmers, pruners, and herders, at least. Each of these is a craft specifically adapted to the trees of the *dehesa*. Careless barking kills the trees, as does bad or excessive pruning. Plowing too close to the trunks will likewise send the trees into decline; shallow plowing farther away will stimulate their growth. Too many cattle, sheep, or hogs will degrade the soil and destroy the forest; the right amount will keep the soil in good tilth and renew its fertility.

In some villages, the pig pasture also remained a common holding, and one man was hired to tend everybody's hogs. The swineherd would pass each house in the morning, calling the pigs from their resting places for a day among the *dehesa's* oaks. In most old Spanish village houses, the livestock occupy the bottom floor. Even in elevator buildings in cities, the ground floor is often called the *bajo,* the beneath, and the *primer plano,* or first floor, is really on the second story of the building.

There was often no controlling the herd in the evening, when they once more came in sight of home. A nineteenth-century English traveler recalled the pigs coming into town "at a full gallop, like a legion possessed with devils, in a handicap for home, into which each single pig turns, never making a mistake." Something similar happened to Don Quixote, when six hundred running hogs came straight through his and Sancho's campsite, upsetting both men, their horse and ass, and all their baggage, including the don's clanking armor.

In recent years, however, scientific forestry and agriculture have reached even into Extremadura. Studies show that pigs fatten quicker if kept in stalls and fed grain. Eucalyptus and poplar thrive on the *dehesa*'s soils. They grow fast and straight, and can be sold immediately to the sawmills. To make these improvements, all you have to do is to cut down and grub out the oaks.

Out of Airy Nothing

. . . and gives to airy nothing
A local habitation and a name.

— WILLIAM SHAKESPEARE,
A Midsummer Night's Dream,
Act V, Scene i

Every moment of every day for several hundreds of millions of years, trees have been creating the two essentials of human life. They mine the air for carbon dioxide, and out of this airy nothing they make both the free oxygen that we breathe and the solid fuel by which we live. One acre of healthy oak woodland takes two tons of carbon from the atmosphere each year. This adds about twenty cubic meters of wood to that acre of oaks, the equivalent of about nine cords of wood. The caloric content of this wood is roughly 234,000 British thermal units.

Fire made human civilization possible. It heated the homes, cooked the foods, made the pottery to contain the grains. But about eight thousand years ago, people learned to further distill the energy in wood, volatilizing the other substances out of the mass and creating lumps of almost pure carbon.

Charcoal was the fuel that ended the Stone Age, for with it one can smelt bronze and found iron. Later, it became a necessary material for glassmaking and gunpowder, as well as a drying agent for the hops and barley that make beer. It has many advantages over raw wood: It is lighter, more efficient, and burns hotter. It is smokeless and easier to control. Charcoal has almost three times the energy content of raw wood per unit of weight, though for most of human history, it took about eight pounds of wood to make one pound of charcoal.

The trick was to burn a pile of wood very slowly, severely restricting the amount of oxygen in the pile. This was sometimes done in a pit in the earth, which is a natural oxygen restrictor. Just before the advent of coal, coke, and oil, charcoal was made in brick or metal kilns, but for eight millennia it was made by colliers in the woods who created large concentric stacks of wood, then covered them with earth to control the burn.

It was a warm-season job. The collier felled the oaks in early spring, cut them into manageable logs, about fireplace size, and left them to season. Usually, the bark was removed to use in tanning before the logs were chopped.

In high summer, the collier would kiss his family good-bye and go live in the woods. He might spend a few weeks there, harvesting a few thousand bushels of charcoal, enough to supply the local ironworkers, blacksmiths, and neighbors who did their own smithing or who used charcoal for heat. Or he might spend the whole summer in the woods, living in a lean-to made of branches, leaves, and dirt, a structure closely resembling the piles he burned.

To make each pile, the collier began by constructing a fagan, a central chimney. Often, the fagan pole, a long column of wood with a cross handle at the top, was his most prized tool and would lean against his hovel when not in use. He propped the fagan pole with oak logs for the burn, building them up into a chimney twelve feet high or more. Around these, he placed whorls of concentrically placed logs until he had a broad conical pile as wide as necessary for the wood to find repose in the pile. A good-sized pile might contain fifteen to twenty cords. Next, he covered the pile all around with soil, making sure to have extra to hand in case the burning pile showed signs of "breaking out," of bursting into flames. Finally, he withdrew the fagan pole and threw burning charcoals down the chimney.

Burning is an exothermic process. It takes heat to start, but it generates more heat than it absorbs. A very slow burn converts wood into charcoal without flame. The heat causes the noncarbon elements in the wood to vaporize and escape into the air. First goes the moisture, then the creosote and tar, and finally the minor volatiles. Black lumps are left, but unlike the highly processed charcoal that we buy today for barbecuing, they still show the whole ring and ray structure of the oak wood that was their origin.

Complete burns took two to fifteen days per pile—depending on size and climate—and a full-time collier might have fifteen or twenty piles going at once. Soil had to be reapplied to keep the pile from catching fire, and as the distillation continued, the collier had to "jump the pile," leaping around the mound to readjust the internal tensions and insure an even burn. A charcoal-making field must have looked a lot like Hell, red glows on black mounds, with the occasional jumping demon.

Few innovations ever changed the process. Thomas Jefferson, who tried to improve on every craft, started lighting the pile from the bottom by inserting a long brand at the bottom of the fagan. This arguably made the burn quicker and more efficient, because the burn would always go from the bottom up.

It is hard to remember that until the middle of the nineteenth century, the collier's work was not a marginal occupation but a central act of human culture. Without the collier, there was no man had his sword—not Sargon of Assyria or Alexander the Great or Charlemagne and Beowulf, not Henry V, not Hernan Cortez, not even King Arthur. Without the collier, no Viking had his longboat, nor castellan his cannon. Without the collier, Notre Dame had no bells nor rood screens, nor did the doors have their hinges. Without the collier, there was no glass for windows, no iron for pots, no making of beer, no founding of wine bottles. Without the collier, there was no gunpowder, nor coins of silver or gold. Without the collier, there was no plowshare, no harness. Without the collier, no ring, no necklace, no crown.

Temper

Come, let us obey the creative word,
God will make us flash like the blade of a sword.

—HUGH MACDIARMID

Steel is where the mine meets the forest. A great sword is not made of iron alone, but of iron wedded to carbon and led through a process of refinement. Plato wrote that all human industry had its source either in mining or forestry. Swordmaking is derived from both.

Oak charcoal—preferred because of its high heat content and steady burn—fueled the smelter in which iron ores were refined. Often, the roaster was filled with alternate layers of charcoal and iron ore. At about seven hundred degrees Farhenheit, the iron simply became malleable, giving up to the air any carbon that had been in the ore. The result of this smelting was pure wrought iron. It was very flexible and could be hammered into intricate shapes. Unfortunately, it was too soft to hold an edge, and it bent too readily to be a sword.

Doubling the temperature in the forge—by introducing forced air with a bellows—changes the direction of carbon flow. At that point, carbon from the charcoal migrates into the iron's loosened crystalline structure. Eventually, the iron liquefies and can be poured. This is cast iron. It has the virtue of being quite hard, but it is also brittle and inflexible.

The great art of the swordsmith and of all the fine toolmakers was to learn how to regulate this process so as to yield steel, a material that is both hard enough to hold an edge and flexible enough to stand up to repeated shocks and blows. We know now that these tool and sword steels have ideally between 0.4 and 1.0 percent carbon in their structures, but the earlier smiths judged their work by the color of the hot iron.

To make steel, the smith first heated the iron until it was cherry red, signaling a temperature at which it had taken up the maximum amount of carbon, and then quenched the iron quickly in warm brine or oil to create hardened steel. (Freshwater was seldom used, because the bubbles that were generated as the water boiled, gathering around the iron, could cause the metal to crack.) The exterior of a bar of hardened steel would be hard, the interior malleable. The hardened billet would take a fine edge, but it chipped easily.

Sword and tool steel had to be tempered. This was where the greatest skill was needed. The smith reheated the metal slowly, noting carefully the colors that it turned. Hardest was straw yellow or bronze. Purple and blue were intermediate. Dark blue was malleable enough to make springs. When it had the color he desired, the smith removed the steel from the heat and cooled it again in water. The effect of tempering was to relieve some of the internal stresses of the carbon-iron matrix while leaving enough intact to permit the sword to hold its edge. Too much heat and the carbon migrated out of the iron. The steel becomes annealed, that is, reduced to its original softness.

Many blades have been ruined over the centuries by zealous grinding, for exactly this reason. Regrind the bevel of a blade on a rotary grinder—even more so on today's electric grinders—

and the blade heats up as it is applied to the gritty stone. Too much heat for too long and the carbon-iron matrix is destroyed. The blade loses its temper.

The smithy is the origin of the phrase "to lose your temper," which we use to describe a person overcome by anger. It is a peculiarly apt phrase. It is not the same as being dull, stupid, or silly. Too much heat, too much anger, and you put your whole being into a kind of stupor from which recovery is more difficult than it seems.

We like to think that our times are much more precise than were times past. Yet as a description of a psychological state, it is hard to beat the metaphor of "losing your temper." There is an exhaustion to that state out of all proportion to the amount of energy exerted. It is as if the quality of anger, self-righteousness, indignation removes an essential element from the mixture of one's life, leaving one limp and defenseless. In such a state, merely trying to buck up or get back on track often will not work. Something is missing in the whole mix of the being. It has to be acquired again. If one knew the work it took to make one's blade hold an edge, and the danger one ran by overheating it, one might be slower to give in to anger.

The Impossibility of Dolphins

How can dolphins swim so fast? They are often seen alongside speeding ships, passing them, and diving and reappearing on the other side.

In 1936, the British zoologist Sir James Gray calculated that in order to maintain such speeds, the average six-foot-long dolphin requires about 240 pounds of muscle. But the cetacean has only 40 pounds at most. The dolphin is manifestly impossible, yet it exists.

The U.S. Navy wanted to find out how this could be. If they could learn the secret, they could design faster ships and torpedoes.

Gray suggested that the way a dolphin moved through the water had something to do with its ability. A dolphin does not simply meet oncoming waves like a sharpened beam; rather, its shape constantly adjusts to present less resistance to water and wind. In short, the dolphin wiggles.

The Navy never managed to make a steel ship wiggle, but if they had wanted to see a successful realization of the principle, they needed only to study Viking ships.

Motion through moving fluids is widely acknowledged as the most formidable and complex of all engineering problems. The surface of the sea is an infinitely variable wave system. (The sea is in almost every culture a realm belonging to the divine, because it is absolutely beyond our power to predict or control.) And air is not only infinitely variable, it is also invisible and weightless, even more so a part of the Creator's invisible world that enfolds and makes possible our own.

It is understandable that the Navy might have made a mistake, not just because the problem is a hard one, but also because modern ships are made of steel. They *do* behave like beams. Twelve centuries ago, however, the people of northern Europe created boats of cleft oak that behaved very much like living dolphins.

Ole Crumlin-Pedersen, who has spent his life studying, resurrecting, rebuilding, and sailing ships of the Viking era, remembers the first time he rode in one. It was the *Roar Age*, a ship painstakingly reconstructed using ancient techniques. The dimensions were derived from those of the first of six boat finds that had been made at the bottom of Roskilde Fjord and which would form the raison d'etre of Roskilde's wonderful Viking Ship Museum. "It was uncanny," he remembers. "When you stood in the bow, the whole ship was moving around you. You could feel the planks moving under your feet, and you could see the gunwales twist and the prow rise." The same thing happened in the stern as the parted seas closed around the passing ship. Only amidships was the oak still.

Magnus Andersen, who captained a replica of the Gokstad ship, the first intact Viking ship ever found, and sailed it to the 1892 Chicago Exposition across the open Atlantic, reported a similar experience. The gunwales, he noted, would twist as much as sixty degrees out of horizontal, but the ship did not leak, and it made twelve knots without straining.

Pound for pound, no better boat than the northern longship has ever been built. The mature Viking ship could make up to twenty knots under its square sail, five knots rowing. The crew could hoist the sail to make better speed or to terrify an enemy. They could drop the sail and start rowing, in order to arrive unheralded. Broader versions were made for trading, narrower ones for raiding.

By means of this ship, the Vikings conquered Ireland and part of England, inhabited and colonized Scotland, the Hebrides, the Orkneys, the Faroes, and the Shetlands, as well as Iceland and Greenland. The Vikings founded both Dublin and York, and ruled in the east of England for better than one hundred years. They reached the coast of North America half a millennium before Columbus. They established the kingdom of Normandy (literally, the Northmen's Land) in France, and sailed down the Seine to harry Paris. Anyone whose ancestry is in the British Isles likely has a double shot of Norse blood, first from the original invaders and colonizers, second from the Normans who conquered Anglo-Saxon England in the eleventh century. Viking traders reached Rome by sailing coastwise into the Mediterranean, and Constantinople by sailing down the Volga and portaging to the Black Sea.

With its slender, narrow, and shaped oak strakes, the long-ship created a standard for movement over water that has never been surpassed. Under sail or oar, it could turn on a penny, and it drew only about three feet of water, meaning that it could sail up all but the shallowest rivers. It was nevertheless strong and durable, and it was light enough that it could be drawn up onto the beach by its own crew. The average ship lasted twenty years and some were in service for as long as a century.

Not every one of these ships was made of oak. On the Atlantic coast of Norway north of Bergen there were no oaks, so the people had to make do with fir. But throughout Europe, wherever there was oak, the ships were made of oak. More than 90 percent of the northern ships were oak. Without oak neither the Vikings nor the later maritime nations of modern Europe would have been able to cross the seas or circle the globe.

Oak was the first choice for shipbuilding because it was strong, comparatively light, watertight, bendable, and, above all, workable. Oak could reliably be split not only from one side of a log to the other, but also from the edge of the log to its center, following the natural lines of the ray cells. Because it had been split edge to center, however, the cleft board had a profile that was slightly wedge-shaped.

At first, this must have been regarded as a drawback, but sometime during the fifth or sixth century A.D., a builder suddenly saw it as a revolution: If you could overlap the boards—the narrow edge of one over the thick edge of another—and securely join them with iron rivets so that they did not leak, then you could bend them into a ship's shape and make a light, streamlined object that would swim through the water like a fish.

The people who made the longships could neither read nor write. Their units of measure were not the degree and the millimeter but the thumb (inch), the hand, the cubit (elbow to fingertip), the foot, the ell (shoulder to wrist), and the fathom (outstretched fingertip to fingertip). Yet they made thousands upon thousands of these boats in an unbroken tradition now more than fifteen hundred years old.

Oak was worked four ways to make the boat. The keel was always taken from one or more tall slender boles, wood as knotless as possible. The mixed hemiboreal forests of Scandinavia yielded oaks perfect for this purpose, since they had to grow fast and with few lower branches in the dense, partly evergreen forests. The trunks would be squared with an ax, then given a wedge or a rabbeted shape by reduction with ax and adze. The finest keels were made of a single trunk. The Gokstad ship, for example, had a keel almost fifty-eight feet long. Lesser ships might have their keels composed of two or three oaks, carefully scarfed together.

The planks for the side strakes were split out of whole logs. If the grain ran straight, and if he were careful to balance the tension on the different wedges so that the grain did not run out or tear, the woodcutter could make planks as long as eighteen or twenty feet, though the average planks were twelve feet or less. Anyone who has ever split oak will remember with deep satisfaction how a crack will open along the length of the bole and keep opening straight and true as the wedge goes deeper, making a pleasant cracking sound. A trunk three feet in diameter, split by an experienced craftsman, could yield up to twenty usable boards. No wood was sawed in Europe before the twelfth century, and for the Viking ships this was a good thing, since a cloven oak board one inch thick is as strong as a sawn board twice that thick.

Almost every Viking ship had a strong plank at the waterline, the *meginhufr,* that served, along with the keel, to stiffen the ship.

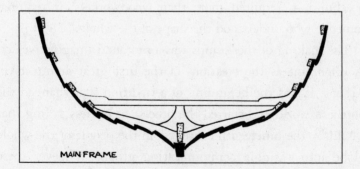

Shell construction of a Norse ship (Nora H. Logan, after the
Viking Ship Museum, Roskilde, Denmark)

From the outside it looked like all the other planks, but it was almost twice as thick. It was not made by splitting, but by gradually reducing a half log to the desired shape and thickness. Axes and adzes were used to do the shaping. A lip might be left on the inside to provide a surface against which to place the knees and the other parts of the light skeleton that was inserted after the shell was finished to help the ship hold its shape.

The skeleton came last, not first, to the longship. The frame was fit into the finished shell and lashed there or attached with wooden nails called "trenails." The purpose of the frame is to spread harmlessly any stresses that impinge upon the hull from outside, and it was natural to use for the framing the trunk-to-branch crotches of the oak, which had grown to perform the same function in the tree. Shaped like broad Y's, these crotches were selected for the purpose in the field, chopped out of the parent trees, and reduced with the ax.

Every Viking household had skilled craftsmen who could choose, cut, and finish oak. The ax and the adze, the mallet and the wedge, were cherished possessions of every family. But boatbuilding required more than woodworkers; it required someone who understood the shape of the whole.

The making of these ships was a profound imaginative act, as important as the creations of the first great architects in Athens. It was the beginning of a tradition that changed the northern world and started the process of restless sailing that would by the nineteenth century join the people of the whole globe into a single communicating group. Whoever these builders were, they were craftsmen as important to the north as the architect was to Greece and Rome.

The northern boatbuilder was called a "stemsmith," because the stern, or prow, was the key element in the design. Building always began with the stem and the keel. Using a knotted string, the stemsmith measured onethird of the keel length. Then he traced a circle upwards from a certain place on the keel using that measurement as the circle's radius. That basic circle contained the shape of the stem. He then carved the stem from a single piece of oak. Sometimes the stem was simply a curved piece of wood into which the planking would be rabbeted, but for the better boats it was a three-dimensionally carved piece of extraordinary beauty. The carving specified how the planks, both larboard and starboard, would run.

More than anything else, the stem resembles a hawk or an eagle about to leap into the air. It has a sharp rising forward edge, but behind this edge trail two wings incised with the angles for the attachment of the planks. The stemsmiths must have envisioned the finished shape as they made these,

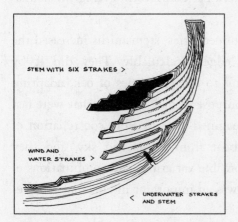

Stem of a Norse ship (Nora H. Logan, after Eric McKee)

willing the water to pass down the sweeping curves of the strakes.

Once this basic shape—keel and stem—was assembled, the stemsmith attached a line from stem to stern and measured the proportions of each plank and rib, using a system of radii marked on a knotted line. The system was plain enough that all intelligent craftsmen could follow it, but the stemsmith had to be there to set it in motion. A failure of his imagination would result in a faulty boat, and a few minutes in the northern waters were enough to drown anyone.

A craftsman's intellect is valuable because it is tested. A poorly made barrel will leak. A poorly made doorjamb will tilt. A poorly made chair will not support. A poorly made boat won't float.

Or else it will be cranky. The word *cranky* now usually refers to someone irascible, unstable, unreliable, but it is derived from boatbuilding. If the craftsman got the center of gravity of the boat too high, it would tend to wallow and be slow to right itself. Instead, it would jerk back and forth, giving the helmsman endless trouble.

In little more than four centuries, stemsmiths increased the depth of human knowledge incalculably. They did it by observing and responding to the properties of oak, adapting these properties to the purpose of making ships that were fast and seaworthy. They imagined the invisible constellation of forces that beat on a boat from water and sky, and they designed for the unfathomable variations and combinations of wind and weather. This was an important intellectual achievement for human beings.

As the oak boats went faster, the stemsmiths' knowledge had to deepen. They learned, firsthand, the laws of motion, by experiencing the results.

Any modern student of hydrodynamics can tell you that as a boat gathers way, it sinks down in the water. Both the bow and stern waves resist the boat's forward motion. At a certain speed the trough created by the bow wave has the same length as the length of the boat. At that point, the ship squats very low in the water. Then, if more force is applied, the ship rises up and begins to plane.

The stemsmiths learned this law the hard way, as did the men who reconstructed their ancient boats. Arne Emil Christensen, the Norwegian scholar of Viking craft, found out about it when he sailed a faithfully built replica of the Oseberg ship, a very broad longship from the ninth century, when it served as the boat of a magnificent ship burial (see page 144).

In 1988, Christensen took the replica out into the Oslo Fjord for a trial. They raised the sail and soon the ship was doing ten knots with ten degrees of heel in a freshening breeze. "At that speed," remembers Christensen, "the ship should begin to rise up on its own bow wave and gather more speed." He sighs. "Unfortunately, at that point the water began to come in over the bow." He frowns, remembering it. "Very fast." The ship sailed itself straight to the bottom and had to be rescued by tugboats. Later, in order to sail it at speeds above ten knots, the crew had to install a makeshift system of shutters that could be raised to repel water at the bows. Finally, the design was modified by adding a number of additional strakes.

The stemsmiths must have faced exactly the same problem,

and they responded by adapting the ships' shape. At the Viking Ship Museum in Oslo, you can stand in the center of the cruciform building and look straight at the Oseberg ship on one side and the Gokstad ship on the other. The Gokstad was built about seventy-five years later than the Oseberg. It has much smoother lines and less belly than the older ship. The Oseberg really looks like two ships: a steep-walled narrow ship built atop a broad-bellied wide ship, the extra-thick *meginhufr* placed midway from keel to gunwale marking the transition. The Gokstad is not narrow and it too has a *meginhufr,* but the transition zone between bottom and superstructure is almost seamless.

Until two hundred years ago, *intellectual* did not mean abstract. It meant the fullest use of all the human faculties in pursuit of the truth. This is what Thomas Aquinas was talking about when he wrote that the highest forms of generation are the most internal. Rocks, he said, evolve only from the outside, by things rubbing against other things. Plants, on the other hand, generate their own successors, but they do this by making a fruit that must drop off and reproduce on its own. Higher animals, he wrote, generate their young internally, but still give birth to them into an outer and separate world. The angels do something still better, though I am not sure I understand what. God does the best because there is no difference between his knowing and his doing. God's work, to Aquinas, is the most intellectual, by which he appears to mean the most creative, for by it God generates and sustains the world.

For God, the connection of inner thought to outer realization is intrinsic, immediate, eternal. In shaping an outer object from his inner imagination, the stemsmith emulates the Creator, because he insists on realizing what his inner eye envisions, connecting the inner to the outer. Intellectual knowledge is not abstract knowledge. An intellectual is a person who puts reflection into action.

The Boat and the Grave

The stemsmith had to imagine the invisible, to design for the unforeseeable. The fact that the people of the north could make a boat that sailed the open sea gave to members of those cultures more confident access than ever before to thoughts of the unknown, the unpredictable, and the incalculable. Reflection grew deeper and indeed stranger, because the boat encouraged people to imagine what the unknown was like and to wonder what their place in it was.

To the northern peoples, the one thing as unknowable, unpredictable, and inevitable as the sea was the certain death—sooner or later, violently or peacefully—of every single human individual. The oak boat gave them new ways to enact the faith or the hope that the dead have not gone out of existence. It became a central part of the dramas by which memorials were created.

To the coastal peoples of the Viking age, the world had no edges. Beyond the known was the West, the sea. Boats of oak were the emblem of possibility. Boats brought wealth from abroad. They took people to new lands and new homes. They brought in fish off the sea and grain from the market. They took men to win fame or to die trying. Oaths were sworn while touching a boat.

A Faroes Island proverb goes, "Bound is the boatless man." A person in a boat is unbound. A person in a boat is going somewhere. A panegyric verse to King Canute reads, "The wind is blowing you, high ship chief, to the west. You print your name in the sea with the wake of your stems."

In boat burial, the people sent their dead into the water, into the wind, and even into the mind. Each choice was a dramatic speculation on the character of the unknown. In *Beowulf,* for example, the dead chief of the Swedish Schieldings is laid out in the middle of his own boat, his weapons piled about him, and the boat is pushed out to sea while the people wail. Likewise, in literary retellings such as Njal's Saga and the Arthurian stories, creatures come in boats from over the sea or from the middle of a lake to bear way the beloved dead, arrayed for his or her future life.

The boat and the dead might also be committed to the air. The only actual record of a boat used in a funeral pyre is from the journal of an Arab diplomat, Ibn Faldun, who witnessed it on the Volga River in the year A.D. 920. The ship chief was laid out on an elaborate bed dressed in Byzantine brocade. Wood was piled around the ship, which had been pulled up onto a raised mound. A servant girl volunteered to accompany her master and was herself dressed in finery and given plenty to drink. She then had intercourse with all her master's retainers, each of whom said to her, "I do this in memory of your master. When you see him, please tell him so for me." She was then killed, laid beside her master, and the whole ship was burned to ashes.

However the visitor may have sensationalized the facts, it is clear that the ship was the dead man's way to whatever elsewhere he was going, and that he set sail by going up in smoke.

Many funeral ships were laid in the earth or in the beach sands, the dead travelers supplied with their precious goods. In one such burial, found on the edge of a beach at the northern tip of the Orkney Islands, a man, an old woman, and a child were interred together in an oak boat facing the North Atlantic. The man had his sword, his arrows, and a bag of game pieces. The woman had her loom tools, her shears, her sickle, a gilded brooch, and a beautiful whalebone plaque showing a symmetrical pair of dragonlike heads.

We cannot now read the meaning of such objects, nor can we reconstruct the funeral itself. Burial rites are one of the origins of theater, so it is likely that the acts were more important than the props. But just as you can hear the music of a poem even when you don't understand the language in which it is written, so the grave goods in the ship tell a story whose resonance reaches the heart. Underneath it all, the ship says, "I travel."

In one burial ship, however, the dead were buried with what amounts to a map of their journey into the unknown. Discovered in a mound in 1904, the Oseberg burial consisted of an oak ship more than seventy-two feet long and sixteen feet wide, with a burial chamber amidships. The whole ship was tethered to an underground boulder. Fifteen horses, four dogs, and two oxen had been sacrificed and placed in the bow. Also on the ship were a highly decorated wagon, two sledges, a series of animal-head posts, and a chest containing staffs, lamps, and other regalia. There was also a small loom and several long panels of figured tapestries.

The Oseberg ship was not the burial of a great warrior. It was the resting place of two women: one about sixty years old and one about thirty. It is likely that one or both of these women wove the tapestries found in the ship on the very loom found there. Furthermore, the tapestries depict a procession or a series of events that appears to involve not only the women themselves, but also almost all the grave goods found on the boat. There are horses drawing wagons like the one interred; there is a woman holding a pole lamp like one found in the iron-bound oak chest; there is a driver using a staff very much like another object found in the chest.

There is a depth of shaping imagination here not previously seen in the north. To make a picture is to take an event out of time. This one is a memorial of the acts of these women's lives, yes, but it contains a further question.

In the procession is a wagon like the wagon found on the boat. And hung on that wagon are tapestries like the tapestries found lying in the wagon on the boat. One can imagine, then, an infinite regression of tapestries on wagons disappearing out of time in the depths of the picture. For presumably, that tapestry in the picture also contains a wagon with a tapestry, and so on. The picture weaves a possibility that only the mind can follow. It suggests the reality of invisible, internal life.

There is at least one thing in the tapestries that is not on the boat. There is a forest, recognizable by its woven branches, and on the branches hang dead men. The horse that draws the

wagon containing the women and the tapestries is just entering this forest section.

For what purpose?

The oak woods were a place of renewal, and the wood of oak
itself was likely so conceived. Well into the twentieth century,
it was still believed in many parts of Europe that a hollow, living oak—sometimes called a "drum oak" for the sound it
makes when you hit it—was the home of living spirits. In parts
of Ireland, people turned their coats inside out when passing
such a tree, a charm to repel the oak's denizen. And oak was the
central actor in the annual drama of the death and renewal of
the oak king.

Long before ship burials, the people of the north placed their
dead in oak logs hollowed out for the purpose. A Bronze Age
man was found in a barrow near Gristhorpe, in England's East
Riding, for example. He was six foot two inches tall, and lay in
a fetal position inside his split oak trunk. With him were a
bronze spear, a flint spear, flint arrowheads, and part of a ring.
Atop the coffin were crossed oak branches. Such Bronze Age
oak cysts are the origin of the modern coffin, and even today,
there are many people who can't imagine being buried without
their "suit of oak." Both oak and boat spoke of renewal.

The custom of oak burial and even boat burial was powerful
enough to survive Christianity. Not only did Christians adopt
the practice of burying in oak coffins, but in some northern
churchyards people were buried in boats. The derelict grave-
yard in Sebbersund, Jutland, was associated with a stave church
that had disappeared by A.D. 1100. Half of the almost five hundred graves excavated there show the coffin remains, some

derived from logboats, some from the planks of larger ships, and some from troughs or hollowed logs. But there are no swords, no spindle whorls, no sickles, no personal objects of any kind except for one small silver coin. The Christian dead did not mean to go anywhere. They needed nothing to take along. After all, they were not really dead, they were at rest, waiting for the end of the age.

The Christians in their graves in Sebbersund churchyard all face east, their heads pillowed on turf. I suppose that on the dawn of the Judgment Day, they will leap from their tombs into the rising sun, like pilots ejecting from their burning planes. But in their mounds the Norse sleep in their ship-shape graves. On the last day, it may be that they will not leap up to join the heavenly choir, but slip down to the fjord and sail away into the curdling air before even the angels are aware. The Vikings, it may be, will sing hallelujah by sailing.

The Inward Line of the Heart

On a side road in a sleepy Essex suburb in southeastern England is the ancient Greensted Church, so small that in recent years it has been served only by what is called with fine Anglican decorum a "semi-retired" clergyperson. It has a few claims to fame: They say that St. Edmund's body rested briefy here on his way to burial after his martyrdom at the hands of the Danes in A.D. 870. There is the grave of a thirteenth-century crusader standing like a sentinel near the road's edge. The parish has an herb festival each summer and sells a nice cookbook that shows how to use the herbs. The ladies of the parish have knit wonderful figural covers for the kneelers in the pews, some with birds, some with flames. One of the Tolpuddle Martyrs—a farmworker who'd been sentenced to transportation to Botany Bay after trying to organize a labor union—was married here in 1839, following his pardon by the king.

But there is also something funny about the walls. The little church tower is white clapboard, and the vestry is brick, but the walls of the nave look like upright logs.

It is a palisade, much like what the walls of Seahenge must once have resembled, except that Seahenge was circular and Greensted is resolutely rectilinear. If the former was a monument to the existence of an inner world, the latter gives that inner world direction and purpose.

At Seahenge or even Bargeroosterveld, what you see is what you get. But at Greensted, the apparently simple hengelike walls conceal a building of great sophistication. The structure is not the whole point; the building is *about* something.

The medieval poet Geoffrey de Vinsauf wrote a treatise, *Poetria Nova,* in which he compared the writing of a poem to the architecture of a building.

> If a man has to lay the foundations of a house, he does not set rash hand to the work; the inward line of the heart measures forth the work in advance and the inner man prescribes a definite order of action; the hand of imagination designs the whole before that of the body does so; the pattern is the first prototype, then the tangible. . . . The inner compasses of the mind must encircle the whole quantity of material beforehand.

For the medieval maker, the inner world was the starting point, not the destination. It was a world where ideas met deeper and deeper funds of memory, and where new ideas mixed with old.

Buildings framed out of oak posts and beams were a fine language with which to express this inner world, because they were extremely flexible. Furthermore, you could test them before having to hazard them in the world. Almost all timber-framed buildings first were cut and their frames test-assembled on the ground. Only afterward was the building erected.

The church at Greensted is a physically small but poetically rich example. It is a Christian monument, a rectilinear build-

The church at Greensted-juxta-Ongar
(Courtesy of the Antiquaries Society of London)

ing. The people enter at the west and face the altar to the east. The liturgy leads them symbolically from the kingdom of this world to the kingdom of heaven, tasted at the altar. But they pass along a line of vertical oak trunks, a reminder of the henges, of the oak king, and so of their own ancient past.

Greensted is a pun in oak: logs but not logs, henge but church. It is dark inside—you blink when you enter. But once your eyes adjust, you don't see the logs. You see tall, flat boards. Years of use and abuse have colored them a rich deep brown, in places almost black. Yet when you come close, you look straight into the heart of oak trees that were cut and split one thousand years ago.

Though there is architectural evidence that the original walls were in fact earthfast—that is, they were fixed simply by being set in a trench in the ground, just like the henges—the building is everything but a butted line of upright trunks. Each log—they range in diameter between eight and eighteen

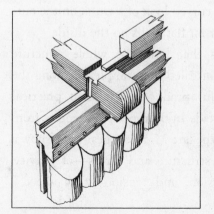

Behind the log facade, a complex of joints (Courtesy of Cecil Hewett, *English Historic Carpentry*, Phillimore Publishing Ltd.)

inches—was split in half, simple enough even for the Neolithic builders of the Sweet Track. But then each was neatly grooved all around the sides. Very slender adzes have been found that might have helped accomplish this work, or it may have been worked by using spoon-bit drills to get the depth and then a chisel to make a smooth, neat groove. Adjoining logs were matched to each other with a slender filet of oak that fit into their matching grooves, creating not only a sound joint but an effective barrier to winter wind as well. At the top, each log was neatly beveled, its top edge formed into a tongue that fit snugly into a long groove in the wall beam above.

Although the church has been much adjusted and rebuilt, what remains also suggests that on the west side the log framework once ran right up to a pitched roof gable. The trunks were not long enough to reach the tallest part of the gable, however, so the carpenters selected a second section of trunk, fit it perfectly, and set it atop the lower log on a tongue-and-groove joint pegged in with oak trenails. Furthermore, to make the corners of the building, they precisely cut a quarter section out of each log, so it would preserve the curved look of a trunk on the outside while making the corner for flat walls on the inside.

The wall and tie beams that hold the whole structure together are likewise well concealed but complex. Among the most elegant and thoughtful products of human practical intelligence are the joints made by European carpenters from the time of Greensted until the late Middle Ages. They show a deep understanding of the structure and forces—of gravity, racking, shearing, compression, and tension—that affect a standing structure.

At Greensted, the challenge was to keep the log walls from leaning right over. The top beams were grooved and slid onto the tongues in the beveled wall logs, holding them all together. So that each of these beams itself would be stable, its ends were fit into a complex joint made at the corner post. Essentially, it was cut so as to trap the end of the plate beam, preventing it from falling, leaning, or shaking loose any of its tongued wall logs. To keep this whole structure steady, tie beams were fixed into grooves cut in the top of the wall beams and held in place by a second beam fit into a lap joint on their top face.

Geoffrey de Vinsauf's analogy between building and poem is more apt than he knew. He meant to admonish poets to give shape to their work, but he inadvertently recognized the poetic function of the framed building. Doggerel or verse, every framed structure was a poem whose language was oak and whose syntax was joints and frames.

Joints and Frames

When a structure was as small as Bargeroosterveld or even Greensted Church, great skill and vision might be involved, but the stakes were comparatively low. Once buildings began to be made of greater sizes—from ample barns and halls to palaces and cathedrals—structural failure could be catastrophic. A scribe recorded the following floor failure in the Anglo-Saxon Chronicle for the year 978: "In this year, all the chief 'witans' of the Angle race fell at Calne from an upper floor, except the holy Archbishop Dunstan, who alone was stayed upon a beam; and some were sorely maimed, and some did not escape with life."

Like the stemsmith, the carpenter had to see what could not be seen. Both men had to imagine and account in their creations for the effects of mass, wind, and storm, but the maker of ships had help from his medium: A boat is cushioned by water from the effects of gravity. It can ride shallow; it can ride deep. It can bend and heel. It can run before the wind. Not so the building. From the moment it is completed, it must stand against the force of gravity, which would have it down. No one has understood gravity so well as the carpenter.

It was no mean thing to be a carpenter in Europe. In the great hall of a king or prelate, the carpenter sat with the priests, rectors, and merchant adventurers. The finest carpenters ate better than that. Hugh Herland, the builder of Westminster Hall, built not only for King Richard but for the king's friend and adviser William of Wykeham, bishop of Winchester. Master Herland often dined with the bishop at his own table, and the bishop had the carpenter's likeness put into the stained-glass windows at Winchester College Chapel.

The carpenter was master at joining one piece of oak to another, in shapes and by means that would resist the worst that weight and weather could do. Between the fourth and the eighteenth centuries in the north of Europe, 95 percent of all buildings were made from oak, cut by hand and shaped into halls, houses, barns, and churches through the use of three principal joints—scarf, mortise-and-tenon, and lap—with dozens of variations and combinations of each.

The carpenters learned the scarf from shipwrights. For a boat to hold a hundred people or thirty tons of cargo, there was not an oak long enough to make the keel or the planks. Likewise, for a hall of any size, it was necessary to achieve lengths of board beyond the scope of any single tree. The solution was to bind two or more beams or planks together, end to end, with a joint that would resist twisting or pulling apart.

A ship like the Nydam boat of the fourth century avoided any scarfs. It was made of a few very long oak pieces. The Sutton Hoo ship of the seventh century had strong, basic scarfs joining keel to stem and keel to stern, as well as plank to plank. The scarfs were made by matching two timbers end to end, cutting long shallow matching inclines into each, and securing the two. Holes were drilled through the joined members and trenails were driven into them.

A pegged scarf was good enough for boats, where the forces were variable but their directions predictable. For a freestanding building, a pegged scarf was less secure. It was only as strong as the trenails that held it together. Shearing or twisting forces caused by a combination of gravity and storm could break the pegs and the scarf would slide apart. When the wind shifts, a boat can change direction; a building can't.

The solutions envisioned and brought into being by medieval carpenters were so beautiful in themselves that it is a shame they were hidden in the unseen joinery of buildings. To prevent the scarf from sliding apart, they cut matching male and female wedges crosswise into the middle of the joint, so that the joined pieces could not slide forward and back. To prevent the joint from racking, or twisting, they cut matching tongues and grooves into each end of the joint, so that they would lock together. And finally, to secure the lock, they cut holes across the joint and inserted wooden keys into them.

To join the end grain of one timber to the side grain of another was the basis of framing and so of house building. Most basically, it allowed the upstanding posts to be joined to the beams that linked them, making quadrangular frames.

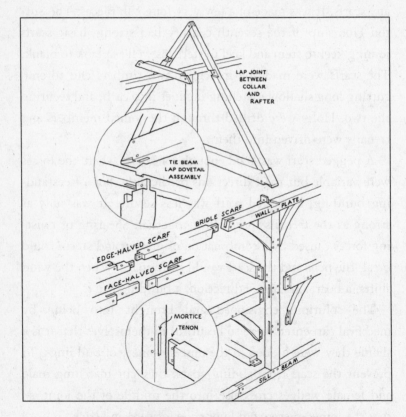

The principal framing joints (Courtesy of Richard Harris, *Discovering Timber Frame Buildings*, Shire Publications)

Wall beams crossed from one post to another along the perimeter of the house. Tie beams crossed from one post to another across the width of the house, creating a series of bays in the interior.

The mortise-and-tenon joint was the key to framing, and its mastery allowed the making of Bargeroosterveld and every post-and-beam building after it. The success of the joint depends on closely matching the male member—the tenon cut into the end grain—with the female member—the mortise cut into the side grain. (Furniture makers use this joint just as much as house carpenters, but the furniture maker tends to cut round mortises and tenons, the house builder rectangular ones.)

The longer and wider the tenon, the better it will resist shearing and bending. If the post tenon is cut inside the full dimensions of the end of the timber, if it is left with "shoulders," the joint will also resist racking, because the shoulders butt against the post and resist bending or twisting.

The lap or half-lap let carpenters connect a long timber across two or more crossing timbers without a break. It also allowed them to match end grain to side grain at an angle, by cutting matching male and female shapes partway through the thickness of each piece so that they overlapped. In some cases, gravity securely held lap joints. In others, where it was possible for forces to separate the members, the angle between the two was cut and notched for each individual piece, to make it difficult for the pieces to twist in a way that would part them. The lap joint connected one large assembly to another—say wall to cross wall or one roof truss to another—to prevent their falling over like a house of cards.

Lap and mortise-and-tenon joints were used together to make the most important joint of all: the complex that held together post, wall beam, tie beam, and roof rafter.

Timber-framed buildings depend on posts to hold up the outside walls. The posts had to be joined to each other by beams to create the frame for filling in walls, and they had to be joined from one side of the house to the other with tie beams to prevent the walls from falling apart once the roof was on. The principal roof rafters had to run from this post-and-beam assembly to the apex of the roof and there be joined to the rafter rising from the other side.

In the earliest timber frames, these joints were usually made separately and in succession. There was a tenon cut into the wall post and a mortise cut into the underside of the wall beam. The tie beam might then be tenoned independently into the wall beam. Once that was assembled, the rafter tenon or lap would be cut and inserted into a mortise or lap on the upper side of the wall beam. The combination was neither efficient nor beautiful. The whole assembly could work and rub in the wind until one of the joints failed.

The carpenters imagined and created composite joints to strengthen the connection. These were visionary acts every bit as profound as the composing of great songs or the framing of great laws. The meditation required to see the joint—how large it was to be, where each part was to be cut, in what succession, with what space between, at what angles—was one thing. But the carpenter had to prove his imagination true by actually making the joint, fitting it together, and leaving the structure to the mercy of the world.

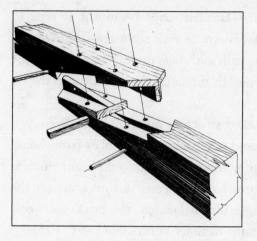

A great thirteenth-century scarf joint (Courtesy of Cecil Hewett, *English Historic Carpentry*, Phillimore Publishing Ltd.)

Memory, reason, and skill are the trinity from which human truth is made. Skill is the doing itself. Anyone who has ever tried knows how hard it is to hollow out a mortise and to carve a tight-fitting tenon by hand. A gouge in the mortise line and the piece will wobble. A hidden knot in the tenon piece and the wood cracks. It is hard for an ordinary carpenter even to conceive the patience, persistence, and delicacy with which the great joints were made.

These joints often required that the post end be left a little thicker to give room to cut the joint out. In some places these joints came to be known as jowls, since they were big and thick like the Hapsburg jaw. The variations were numerous, but the idea was to join the three members together into one solid assembly. Two tenons and a lap had to be cut into the post. One tenon and the lap fit the post into the wall beam. The second tenon fit the post directly into a mortise in the tie beam. The tie beam itself was fit into the wall beam with a dovetail lap joint. This shaped lap—familiar today because it is used to assemble the sides of drawers in fine cabinetry—resisted the tendency of the heavy walls and roofs to pull the tie beam out of place. The rafter tenon fit into a mortise carved atop the tie beam.

The three joints locked structure into buildings. Were you strong enough, you could lift an assembled timber-frame house and put it down again, without any harm to the frame. In fact, when in 1944 a German rocket scored a direct hit on the sixteenth-century Staple Inn in London, the brick walls collapsed but the ornate hammer-beam roof was virtually unharmed.

Carpenter's Mind

The greatest work of art of the European Middle Ages is not a painting, not a sculpture, not a cathedral. It is 660 tons of oak, the timber-framed roof of Westminster Hall. Hugh Herland made it for Richard II between A.D. 1393 and 1397. For the six centuries since, architects, scholars, engineers, and archaeologists have argued about how he did it.

The roof of any building was a carpenter's test and the place he could best show his imaginative understanding of invisible forces. Composed of rafters, collars, purlins, and tiles, the roof was usually the heaviest part of the building not pegged directly to the ground. It exerted the strongest forces, both straight down and outward. And it was always the part most exposed to wind and storm.

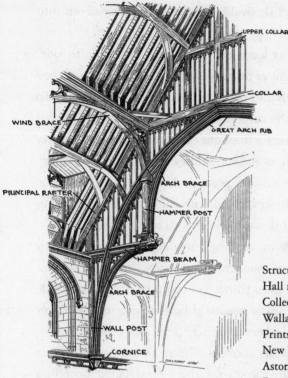

UPPER COLLAR

COLLAR

WIND BRACE

GREAT ARCH RIB

ARCH BRACE

PRINCIPAL RAFTER

HAMMER POST

HAMMER BEAM

ARCH BRACE

WALL POST

CORNICE

Structure of the Westminster
Hall roof (Art & Architecture
Collection, Miriam and Ira D.
Wallach Division of Art,
Prints and Photographs, The
New York Public Library.
Astor, Lenox and Tilden
Foundations)

Each member of a roof truss—the triangular or spire assembly of rafters and tie beams—is a force made visible. Gravity
flows down the roof rafters and pushes at the walls on which
the roof sits. The tie beam resists, holding the two pieces in
tension. Above the tie beam, the carpenter places a collar—a
smaller and higher version of the tie beam—to siphon off
some of gravity's pull. Beneath the collar, a post or a pair of
arched braces would let the forces cascade down to the center
of the tie beam. Beneath the tie beam, more arched braces led

out to a lower level of the walls, letting the forces run out into the ground.

Each truss bears its load, but it is weak from side to side. Typically three to eight or more trusses made up a roof, and they could and occasionally did fall over, landing in a rough pile. To resist the weather, the carpenter ties long, sometimes scarfed purlins among the trusses from one end to the other, sometimes trenched into the rafters and sometimes clasped against the collars. Between tie beams and posts and rafters and purlins, he places arched wind braces, to take up the force of the blowing wind and transmit it down the beams to the ground.

As buildings grew in size, the knowledge necessary to keep them standing became more crucial and more difficult. To span a width of twenty or even thirty feet could be done with single timbers, but larger spans needed different solutions. Carpenters began to scarf timbers together to get the needed widths, arranging hanging posts above the beam to balance the forces that would try to part the scarf. Others constructed ascending towers of arch work to cradle the rafters. Halls could be made even wider by adding an aisle at each side. The outer part of the aisle constituted the outside wall; the inner side was supported on an arcade of pillar posts.

But no roof was as audacious, as beautiful, or as intelligent as the roof of Westminster Hall. Three years before building commenced, Hugh Herland was gathering oak timbers for the job and bringing them to his framing yard in Farnham, Surrey, thirty-nine miles from the building site.

He was not building from scratch. His assignment was to make a new roof for the Norman hall that had been designed for Edward III. The outer walls had a huge span, sixty-eight feet, but this had been significantly reduced with twenty-foot, arcaded aisles on the sides. The Norman high roof had to span only about twenty-eight feet.

Herland decided that he could span the entire sixty-eight feet, without any aisles. This must have pleased Richard II, who could envision his great state events held in a wide and soaring hall. Just as a church led from the world to the high altar, so the hall would lead from the ordinary people to the king's dais, in a soaring theater of the king's making.

They had some reason to believe that the idea was feasible. For more than a century, aisles had been eliminated on smaller spans by substituting hammer beams for arcade posts. Essentially, this meant knocking out the pillar posts that formed the aisles, beefing up the size of the timbers of the remaining aisle beam and the posts above them, and reinforcing the connection between the wall posts and the new hammer beam by means of arched braces.

The triangle formed by the now-cantilevered beam, the post above it, and the base of the rafter made a very solid base to rest upon a masonry wall. The post bore down on the beam, pushing it toward the ground; the rafter bore obliquely on the stone wall, pushing it outward; the cantilevered beam stood in tension between them. An arched brace beneath helped hold the triangle in place. The whole assembly transmitted the combined forces harmlessly into the ground.

To the carpenter, the hammer beam was really just a more dramatic version of the triangular bases used directly atop walls in buildings such as the now-gone church in Limpenhoe,

Norfolk. There, where the steep-pitched roof met the high stone walls, each truss rested on a right triangle of oak. A post and a sole plate (a piece set directly on top of the wall) formed the right angles, and the bottom of the rafter the hypotenuse.

Now, two relatives of the king, John of Gaunt and John Holland, were both having their halls made with hammer-beam roofs at the same time, but the grandest hall of the previous generation had used a different principle. The original roof of the St. George's Hall in Windsor Castle, built for Edward III and probably designed by Hugh's father, William Herland, used grand oak arches for support. The trusses were supported by symmetrical arches that rose to the collar beam.

Hugh had been a subsidiary carpenter on the Windsor project. If Richard wanted big arches to outdo his predecessor, Hugh Herland had not only the knowledge but an extra motive to succeed. By making arches at Westminster, he could not only please his patron, but also honor his father and his training.

None of these precedents, however, could change the fact that no one had ever before spanned a distance anything like sixty-eight feet without the use of intermediate posts. The proposed roof for Westminster Hall was almost twice as wide as its nearest rival.

As Hugh Herland conceived it, the roof appeared to use both hammer beam and arch to solve the problem, while it actually depended on triangles as transmitters of force.

The carpenter had no oak big enough to form the rafters out of single pieces. The massive principal rafters—thirteen pair of them for the twelve bays of the hall—each had a cross section of twelve by seventeen inches. This meant that arches alone would be insufficient to support them, since the scarf joint of those massive timbers would tend to bend and part were they to rest upon curved members. Large hammer beams and hammer posts, joined to the bottom section of each rafter, could hold up that rafter's heavy base, but what was to hold up the hammer beams? It was with these difficulties in mind that Herland designed and built Westminster Hall's new roof, using both arch and hammer beam.

The whole structure was first laid out in the framing yard at Farnham. No timber-framed building of any scale was ever constructed without first being laid out, jointed, and assembled on the ground. The Roman numerals and other marks still seen on the frames of ancient timber structures told the original carpenters in what order and in what place each piece was to be assembled. Indeed, one reason that carpenters tended to be men of substance was that they had to own a substantial piece of land to store the oak timbers and do the framing.

The first two years of the project were spent cutting the thirteen principal frames and their thousands of joints. Finally, on June 1, 1395, the order went out to bring "300 strong wains" to Farnham to begin the process of transporting the prepared timbers to London. For the next month, the wagons took five loads of oak per day down to Ham at Kingston upon Thames, where it was loaded onto barges to be floated the rest of the way to London.

When the wood was on-site, Herland began to build, probably using the leftover arcade posts and other material from the former roof. First, he set the massive wall posts into slots in the masonry wall. These members were already carved with joints that were to receive the ends of braces and arches, so they resembled the carved stem of a Viking ship. Then, he installed the first braces, supporting them on the old arcade posts so that they would stay in place. Next, he tenoned the hammer beam into the top of the bracket, the hammer post atop the hammer beam, and the lower section of the rafter into both hammer post and hammer beam. Next, the collar beam was installed from one rafter to its opposite, for the first time joining the two halves of the truss. Then, the arches—really an assemblage of many slender core pieces with elaborate attached moldings —were fit into the braces rising to the collar. Above the collar beam went queen posts surmounted by a king post that finally reached the roof peak to form a great isosceles triangle.

All thirteen of the great frames were so assembled. Purlins and an elaborate series of arcade braces were placed to connect the frames, to stiffen them against the wind, and to carry the numerous common rafters. In all the frame's voids, slender panel frames were inserted, like long windows with trefoil tops. Angels in flight—both a symbolic support and a hopeful prayer for the soaring structure—were carved onto the ends of the hammer beams.

Then they kicked out the support beams.

For the first time, the roof stood on its own. It has stood so for more than six hundred years since.

Many carpenters must have studied it, but none afterward surpassed it. No sooner had the professions of engineering and architecture emerged out of the Scientific Revolution than engineers and architects began to analyze Westminster Hall's remarkable roof.

One said the roof stood up because of the stone corbels on which the wall posts stand. Others said it survived because of the massive principal rafters. More contended that the hammer beam, acting as a cantilever, was responsible. Some scoffed at this assertion, claiming that the hammer beams were mere decoration. Some said it was the great arches, others the braces, and others both arches and braces at once. Others remarked that the arches weren't really arches, the braces not really braces, and the rafters not really rafters. Engineers did studies, crunched numbers, tested models, and eventually ran computer programs. Still, no one is quite sure why the greatest timber roof in the world stands up.

But intuitively, almost everyone sees that it should. If you envision the forces of gravity and weather acting on one of the trusses, the whole elaborate structure comes to life. Force rolls down the great composite rafters. Where it might harm them—that is, where they are scarfed together—the collar beam stabilizes the members. More force flows straight down from the roof peak into the king post, which in turn distributes its forces to the queen posts and they to the collar beam. The collar beam passes most of its force into the twin hammer posts that are tenoned into its ends. Some force from the collars is passed to the composite arches, which clasp and strengthen the hammer posts. The hammer posts bear down on the hammer beams, which are probably held stiff between this force and the outward thrust of the rafter attached to their other ends. Force then flows off the building into the ground through the top of the wall and through wall posts and corbels.

Seen more plainly, the roof acts as three strong triangles: The first is the isosceles triangle that runs from the roof peak to the collar. The second and third are right triangles formed beneath and beside the collar by the hammer posts, hammer beams, and rafters. But what to do with all the braces and the arches?

Unlike a modern building, the roof of Westminster Hall is not transparent to analysis. It was not made by an analytic mind, but by the mind of a carpenter. What is the difference?

The medieval carpenter did what today's engineer might call "overbuilding." The carpenter called it sound, sensible, right, proper. The carpenter's knowledge came from what he put his hands on, and from materials that he knew both in the woodland and at his yard. He had a long training, in which he probably began with several years merely sharpening tools and cutting joints. He had a deep visual imagination of the way forces act, for he had seen hundreds of buildings—all made of the same oak wood—in hundreds of situations. Each settled, bent, stood, or failed in a particular way.

For the carpenter, prudence, not economy, was the first virtue. He made redundant systems intentionally, much as the oak tree does with its dormant buds and its multiple leaders and its hundreds of miles of roots. A failure of one piece will not bring the whole thing down. The rafters at Westminster Hall are probably too large, and the hammer beams may be too slender. There are braces going in every direction. The arches probably play much less important a role than they seem to, but they do stabilize the hammer beams and hammer posts. When the corbels are removed in a scale-model test of the roof, the whole structure starts to collapse, but the fact is that much of the wood has so decayed over the years that many wall posts no longer even touch the corbels.

No one thing holds up the roof of Westminster Hall. In this, it is like the oak of which it is made: No one thing makes it superior.

Mister Cooper

Watch a wet cooper work. He starts with raw white oak, cleaves it into staves with a froe, cuts the staves to length with a saw, and shapes the staves with a short, broad-bladed ax. He shaves out a concave and a convex side with drawknives of different radii and shaves the square edges of the staves with a jointer into radius lines that will intersect the cask's center. He raises up all the staves around one temporary hoop and adjusts them to fit, then heats the wood with a small fire in the center of the cask, so the wood will bend. He circles and circles the sweet-smelling hot oak, driving on hoop after hoop. He keeps the fire going to set the final shape. Once it cools, he scrapes out the inside of the new cask, either smoothing it for beer or roughing it for wine and whiskey. He cuts and joins the cask ends, scribes a circle in the wood at the chime ends to hold the ends, fits them, and finalizes the hoop. He augers out a bung-hole and fits a stopper. The cask is made.

"There are no amateur barrel-makers," wrote Kenneth Kilby, one of the last of the English coopers, in his memoir, *The Cooper and His Trade*. When you see a cooper work, you know why. Without a ruler, a tape measure, a protractor or any of the tools of our precision age, he turns sticks of wood into a shape of enormous complexity.

If you had never seen a cask, you might not believe in their existence. They seem to be both cubes and spheres at once. Try to imagine a container that holds anything from bourbon to nails to butter, in quantities weighing up to half a ton, without breaking or leaking. It sits firmly on its end and can be stacked five high, yet you can roll it down the road, applying slight pressure to either end to change the direction.

Whoever first discovered this shape was a genius. Johannes Kepler, the seventeenth-century mathematician and astronomer, was moved to calculate the shape of maximum volume for a barrel on the occasion of his second marriage, when he conceived that a wine merchant was trying to cheat him. The merchant calculated the price of the wine by measuring from the bunghole—the hole left for the tap in the middle of the barrel—up to the top of the barrel on the opposite side. "Wait just a minute," thought Kepler. "If the barrel is long and skinny, it could have just the same measurement for this length as one that is shorter and fatter." Yet the second of these would have a far greater volume.

He did the math and found the maximum volume possible. Strangely enough, it was the same volume Austrian wine barrels already had. Kepler realized that coopers had recognized and solved this problem long before he had.

In solid geometry, the barrel shape is defined as "a solid of revolution composed of parallel circular top and bottom with a common axis and a side formed by a smooth curve symmetrical about the midplane." In practice, it is a truncated spheroid made out of many staves. Each stave is a straight-grained, knotless piece of oak, wider in the middle than at the ends and bent to give a "belly" to the assembled cask. If it is not rightly cut, rightly fired, and rightly bent, the barrel either won't fit together or it will end up askew. If your cask leaned, the other men in the shop would tell you you'd made a "lord," as in "drunk as a lord." If it had too little belly and looked practically rectangular, they would tell you, "you're nothing but a carpenter!"—a phrase of insult.

The making of casks was a complex mistery. Not mystery, but "mistery." The boy who wished or was made to learn the trade would likely never have heard of geometry, but after a seven-year apprenticeship, he had become a practical geometer with few equals in his town. He had still not heard of geometry, but he had learned his tools and could make a cask of any volume required, accurately and without leaks. He had earned the right to be called a "mister," or "master" of the trade.

The honorific title "Mister" is a pure (and vanishing) formality. Few people are aware of its derivation. But in the Age of Oak, Mister denoted the master of a craft. It was a powerful honorific, and it existed specifically to distinguish from the other current honorifics: Lord So-and-So, Sir Somebody, the Honorable Diddledee, the Reverend, Very Reverend, Most Awfully Reverend Rubadub. Mister meant that a person had mastered a complex task and could do it reliably and well. It signified a high level of coordination between hand, eye, and brain.

These were the people, argued Thomas Jefferson, out of whom the great democracies were to be made. Not only out of the

Sirs, the Honorables, and the Reverends, but also out of the guilds of town and country. The Misters were men who had trained their intelligences to a high level by encountering and transforming resistant materials. Faced with questions of how to live and how to organize a society, they were, he felt, more to be trusted than those positioned either higher or lower in the social hierarchy. Indeed, the American Revolution was deeply respected in Europe because it was thought to represent a new experiment precisely upon those lines: The Misters were in charge.

In colonial America, where cities were fewer and smaller and where lines of communication were as yet tenuous, the Misters were most often yeomen. Their ideal was to farm and to practice a craft. The land and the farming secured their livelihood and status, while they performed their craft to satisfy the needs of themselves and their neighbors. They learned their crafts the same as any other, but since their customers constituted perhaps a hundred neighbors who lived within five miles of their shop, they could not live by barrels or butter churns alone.

In the larger cities, indeed, there were full-time coopers, housewrights, blacksmiths, potters, but in the country, a man would farm and make casks, farm and raise barns, farm and found tools. Even to this day, when farmers are no longer the leading citizens, the tradition persists of supplementing farm income with a skilled trade. A farmer may also be a machinist, a tractor repairman, a blacksmith, even a banker. The yeoman was a person trained to meet life and therefore a ready, dependable citizen.

Tools were the keys to each trade, and technological change was glacial. The Roman cooper used basically the same tools as the European or American cooper of the mid-nineteenth century. Coopers' tools recovered from a seventh-century Byzantine wreck are identical to those used today in traditional cask making. In one and one-third millennia, then, there had been no change in the tools.

The tools were the medium through which the cooper could meet and shape oak. Often, the first task to which an apprentice would be set was to learn to sharpen. There were tools used to split, to hew, to shave, to hollow, to plane, and to joint. All had blades, some of them with two faces, or bevels, some with one. Some tools were straight, others curved, some reflexed practically into the shape of hoops.

The apprentice would have had to regrind really dull tools on a sandstone wheel, worked with foot pedals and revolving through a bath of water. The water kept the stone and edge cool, so that the blade would not lose its temper. A tool that has lost its temper is almost useless, since it can no longer get or hold an edge (see page 132).

When the grinding was done, or when a sharp tool simply needed to be honed to a perfect edge again, the apprentice used whetstones. For most of human history, he would have lubricated the stone with spit. In the nineteenth century, the finest stones were called oilstones, because they were lubricated with oil. Often, sperm whale oil was specified.

There was a test for sharpness among wet coopers, those whose barrels had to be tight enough to hold wine or beer. They took a piece of the long iris leaves ("flag") used to help seal joints and allowed the edge of the tool to cut along the leaf, using only the tool's own weight. If it made a ragged cut, the blade was not yet sharp enough. One of the first lessons a new apprentice learned was never to try to catch a tool falling from the bench. If it were properly sharp, it would cut his fingers to the bone.

A cooper's apprentice had to sharpen axes, adzes (different shapes for rounding and trimming), augers, borers, bits, bucket shaves, buzzes, cold chisels, chivs, crozes, dowelling bits, downrights, drawshaves, round shaves, flinchers, froes, inside shaves, jibbers, jointers, drawknives, heading knives, hollow knives, topping planes, pluckers, punches, handsaws, frame saws, and swifts. Some called for different files or stones, others for different angles of attack; some had teeth to be filed and others edges. By the time an apprentice could sharpen all of these to the satisfaction of the master, he was the possessor of an intelligence every bit as keen as that of someone able to manipulate the quadratic formula.

The apprentice now had sharp tools, but he had yet to make a single cask. To do so, he had to learn how to convert spherical geometry into plane geometry, and then into spherical geometry again. And not just for one or two sizes. There were more than one hundred different sizes and shapes of barrel made by wet coopers alone. White coopers—the people who made barrels, casks, and churns for milk products—made more than thirty-five different sizes and types, and dry coopers another hundred or so.

The first tool he had to learn was the froe.

A freshly cut oak trunk—called the stick—is a cylinder. The

simplest way to make a stick into rectangular planks is to knock the round edges off, then slice it through and through. This will do for simple applications that don't have to be watertight, but it won't do for barrel staves. Wood shrinks when it is cut. It shrinks from the circumference toward the center along radius lines, but it shrinks twice as much along lines drawn tangentially to the annual rings. If you look at boards cut straight through a log, most of them will show long arcs of annual rings in their end grain and comparatively small sections of rays. As this wood dries, the long arcs pull together much more dramatically than the rays contract. This is why plain-sawn boards or flooring so often warp or cup.

It may be annoying to walk on the uneven surface of cheap, cupped flooring, but it would be more than annoying to find your barrel staves warping and releasing floods of fine wine. The first task of the cooper then is to use his sharpened froe to cleave the trunk radially into staves, which are far less prone to shrinking and warping. The cloven board also repels water and fungi much better than a sawn one. With a mallet, he started the froe into the stick, then carefully but firmly levered his way through the length, splitting the wood. The only use the cooper had for a saw was to cut the boards to length. The next tool the apprentice learned was the broadax. He had to square the board, cutting off the easily rotted sapwood and the bark, and leave the boards to dry, or season. (Fresh, green oak is easy to cut, but it tends to shrink or warp as it dries.) Any boards with knots or internal cracks went for firewood.

Two years later, the staves were ready to dress. The first step was to list them. Using an adze or a small ax with an offset handle, the apprentice would begin to convert the flat board into the gores of a spheroid. He had to cut an angle at each of the four corners lengthwise down the rectangular board, leaving enough width for the belly of the cask and cutting evenly, so that all the surfaces would join without gaps.

Each cask contained at least two dozen staves, and each had to be cut with equal belly and equal edges. The apprentice never stopped to measure these. He cut one stave to his satisfaction, then began on the others, using the first as a model. Not only did he have to make a spheroid that stood straight, but he also had to assure that it would have sufficient belly to hold the correct volume. It was a skill that had to be honed over dozens of attempts before he could get even one size of cask right. Then, of course, he had to learn at least a hundred more specific sizes.

The great master coopers, it was said, could eyeball the whole stave using the ax alone. They could cut not only the proper gore, but also angle the long edges inward so that they matched the invisible radius lines of the future barrel. But such a skill was rare, if not wholly apocryphal.

The usual process for converting the board back into spherical geometry involved three more steps. First, the apprentice would use one drawknife to hollow out the concave inside plane of the future stave. He would next use another drawknife to cut a convex hump along the outside. To perform these operations, he would place the stave on a shaving horse, sit himself down straddling it, and, holding the wood in place with an ingenious foot pedal, work the razor-shape drawknife rapidly toward him. When an experienced cooper does this, the shavings fly off, but a novice has good reason to fear for the safety of his genitals.

The third step in finishing a stave was the most important. It converted the rectilinear sides of the staves into the steep-angled sides that would mimic the radius of the barrel's imaginary sphere, and so allow the dozens of staves to fit tightly together. The proper tool was called a jointer. Essentially, it was a long plane set into a fine and perfectly straight beechwood box. One end of the box sat on the floor, while the other was supported on legs, making a regular inclined plane with a very sharp line right in the middle of it. By pushing the stave at the correct angle over this jointer, the apprentice learned to cut the radius lines.

For a functional barrel, these lines had to be correct to within a fraction of one degree of circumference, but he never used a measuring tool to check. His eye was the check, and the training of his eye was what he suffered the long years of apprenticeship to acquire. It is actually very exciting to watch someone shave a bit of wood off a little board, look down the board with his eye, and then fit it precisely into another board, which has been prepared by the same method. If it was not exactly correct, he reshaved it until it was.

Exactness, not precision, was what the apprentice paid to learn. He learned good technique and honest judgment, since if he tried to fudge on anything, his leaky barrel would tell on him. It was a training of mind and character, as well as of hand and eye.

When the staves were all done and when the apprentice had set them in a temporary hoop to see that they joined properly, he fit the first of the shaping hoops and called "Truss ho!" If there were other coopers in the shop, they would come to help fire the barrel. They placed it over a low, contained fire of shavings collected from the shop floor. The fire softened the wood while the coopers hammered on the hoops of different sizes, one after another, until the cask achieved its proper shape. When it had cooled the cooper had still to cut and fit the ends, finish the inside and outside, and place the proper spigot or bunghole, but the miracle of the spherical cube had been achieved.

At the end of his time, an apprentice in a shop "passed out." (The idiom survives among us to describe someone who is so drunk he grows dizzy and loses consciousness; it may be derived from the apprentice's graduation ceremony.) The apprentice would not know the exact day or hour, but one day, when he called "Truss ho!" the other coopers in the shop would only feign to help him. Instead, they would throw him into his barrel and roll him all over the shop, shouting, and showering him with detritus. Finally, they would tip him out of the barrel at the shop door, and congratulate him on having become his own man, a journeyman cooper. It is quite likely that on more than one occasion, after such rough treatment, the new journeyman passed out after he had passed out.

The cooper's work was not a minor thing in daily life. Wines, liquors, ciders, and beers were all held and aged in casks, and to some extent, they still are. But much else was also held in coopered vessels. Distilling was done in them, and they were used in mining. Clothes were washed in them, and water was hauled and carried. Milk too was carried in them, and butter churned and cheese put down. Pickled and salted meats and vegetables, crackers, flour, and sugar were stored in them; all sorts of hardware was transported in them. Every medieval wife had at least five or six coopered vessels, so there were probably more pieces of the cooper's handiwork in Europe than there were people. Even up until World War I, the United Kingdom produced a million barrels a year for herring alone.

Some barrels were expendable, but some were so finely made they lasted for decades. A good wine barrel might outlive its maker, and the huge hogshead casks in which Worcestershire sauce aged often remained in use for more than seventy years. When Otto von Guericke was working to create the first vacuum, he started by using a barrel, since he knew it could withstand great pressure. Indeed, a wet cooper's barrel can routinely handle thirty pounds per square inch, about the same as an automobile tire. Precious cargo could be consigned to a good barrel. When admirals died at sea, they were typically bound up in a full keg of rum to preserve the body for the trip home. A drunken seamen might confess he had been "tapping the admiral."

Though there were and are machine cooper shops, the process did not lend itself well to mechanization. A machine cooper shop has many different machines, one for each of the different operations. It is more like a woodworking shop than like a factory, because there is no way to standardize the various tasks required of the cooper. The machine simply aids the man, rather than the man surviving only to tend the machine.

Hand and machine coopering survived side-by-side up until the rapid decline of the craft between the two world wars.

Three things killed coopering: bottles, aluminum, and forklifts. Bottles for both wine and beer were rare before the eighteenth century. As they became more common, the demand for casks slackened. Beer and ale do not actually react with the oak—or if they do, the beer sours—so the loss of the cask was no great matter. As to wine, though it acquired a great deal of its character through the interaction with the tannins in the oak, it could still be aged in oak and then shipped and sold in bottles. The rise of aluminum casks in the twentieth century further reduced the need for coopers. Aluminum barrels were tougher and longer lasting than wooden barrels, and they imparted no flavor at all to their contents. Bulk shipment in long, cylindrical metal containers further reduced the need for casks.

But forklifts were the worst, because they removed the very need for barrels. The great thing about casks was that they could be rolled. In order to get the roll, however, you had to sacrifice some product space. With a machine to move the containers, it was actually a disadvantage to make them rollable. Why not use that little bit of extra space for product? So the rectilinear container won, and the barrel died. Only for aging wine and spirits does it survive, because there is no substitute for the flavor that oak imparts to fermented and distilled products.

Still, coopering is now a largely forgotten skill, practiced by a tiny fraction of the number that once called it their craft. The replacements—bottle making, metal cask making, and forklift operating—are shorter on skill and longer on repetition. It takes skill of course to operate the machines, but nothing like the skill it took to be a master cooper. In these cases, the people serve and tend the machines. The machines do the making and the moving.

The great thing about coopering and other crafts was that they occupied the brain, the hand, and the emotions, all at the same time. There was a resistance to the completion of the task, and this had to be overcome by the craftsman's know-how, a composite of his knowing, remembering, and action. Memory, reason, and skill are God's three gifts to human beings, and the simultaneous activity of the three might just be a requirement to become and remain a human being.

Craft is a school of patience. Patience is what you acquire by working again and again on resistant materials. There is never a right or wrong, only a closer and closer approach to wholly useful.

Patience is the mother of joy. It is through patience that we can endure each other's company long enough to love, through patience that we can cooperate in a task, through patience that we can go from abysmally bad to almost all right, through patience that we can restrain ourselves from wasting our lives in anger and disappointment. The patient person waits, listens, expects, hopes, nurtures, cares, remembers, speaks, trusts, and is courteous. The impatient person demands, gets angry, hurries, presumes, is careless, despairs, forgets, complains, distrusts, disrupts.

Mister Tanner

I buy hides and skins and I prepare them by
my craft, and I make of them boots of
various kinds, ankle leathers, shoes,
leather breeches, bottles, bridle-thongs,
flasks and budgets, leather neck-pieces,
spur-leathers, halters, bags and pouches,
and nobody would wish to go through the
winter without my craft.

—a tenth-century shoemaker,
in Aelfric the Grammarian,
Colloquies

Tanners were not known for their subtlety or their refinement. Their job was to take raw animal hides and skins—often complete with hoofs, hair, fat, and gore—and convert them into leather. A tanner's day every day was to pound, scrape, wash, haul, shift, scrub, and stretch wet hides as big as eight feet square and as heavy as bales of hay.

And the work stank. Around the Western world, villages and towns regulated the places where tanners could set up shop, making sure they were not upwind or upstream of any fine residences. The smell of the bloody hides was bad enough, especially as they went rancid, but the tanner would also bag the hair he scraped off, selling the smelly mess to makers of felts, pads, and carpets, or so as to be mixed into plaster. Worst of all, to prepare the hides for tanning, he would soak them in a solution of fermented hen dung.

In spite of all this, the tanner was never ostracized. He was the key member in a chain of "misteries." His leather was the basis for what everyone needed: shoes, sandals, boots, leather bottles, and harnesses. No one could work in winter, on hot sands, or on rough ground without him, nor could the animals plow the fields, nor could the soldiers march to war, nor could liquids be carried in quantities too small for barrels. Mr. Shoemaker, Mr. Saddler, and Mr. Butler (the word *butler* is short for bottle maker), not to mention Mr. Glover and Mr. Bookbinder, needed first the work of Mr. Tanner and his two associates, Mr. Barker and Mr. Currier.

All the leather crafts depended on the tanner, and the tanner depended on the oak. Indeed, the word *tan* is derived from the Latin for oak bark. When the bark was ground fine and soaked in water with the hides, it released tannins that prevented the skin from rotting and that made it supple and virtually waterproof.

No one knows who first made leather or when, but tanning may be among the most ancient of crafts. Some bone scrapers found at Paleolithic sites are the same shape and size as the two-handed, curved unhairing knives used by tanners tens of millennia later, indeed until the early twentieth century.

The tanning process—laborious and smelly—was also complex. The tanners did not know why it worked, but they knew that it did work, and they handed down the methods intact and virtually unchanged from Roman times until the nineteenth century.

It began in the oak groves with the barkers, who went out in spring as the first flush of leaves came on the trees. Sometimes, the tanner barked his own trees—to encourage tanners, some towns granted them rights to a certain number of oaks from the commons each year—but often whole parties of men, women, and children went out as day laborers to do the barking.

The work had to be done quickly, for only during the early spring would the bark pry easily away from the wood. Barking was often almost a holiday, since it meant extra money and a break from the daily routine. After compulsory education was instituted, teachers would sigh in April at the lack of pupils in the classroom. "The attendance is diminishing," wrote one teacher in Wales in 1865. "Children engaged carrying bark in the woods . . . The bark carrying will probably continue a month or more."

The best tanbark came from coppiced oak, grown on a twenty-five- to thirty-year rotation. At about that age, the long straight stems had plenty of tannin in the inner bark without too much rough exterior bark. But any oak, large or small, might be used.

If a huge oak was felled, the barkers swarmed over it with their billhooks, mallets, and barking irons, or spuds. The first of these was a sharp, curved tool with which to cut rings around the trunk and long lines from ring to ring. With the second, some of the barkers would pound the tree to loosen the bark. The third, the iron or "spud," was a tool that is still often found in antique shops in oak country, where people are mystified as to its purpose. It looks like a sharpened iron heart stuck on a rod and finished with a simple handle. Once the billhook had made the cuts and the mallet had loosened the bark, the spud was inserted beneath the bark to pry it away from the trunk.

When the bark came away, it made a noise like a quack, so a party of barkers sounded like a flock of ducks. A good barker could often strip sections of bark six feet long and a foot broad. When they were done, the remaining wood looked as white as bone.

People of all ages joined in the work. The older men and women worked on the ground, either levering out the bark from small felled trees or helping to attack the larger trunks. Young women climbed onto ladders to bark lower branches.

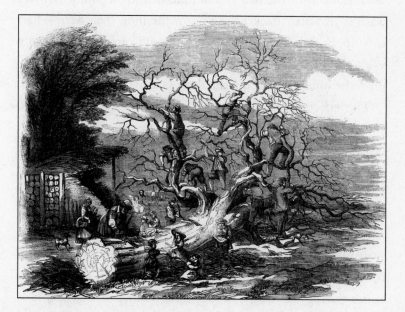

A bark-stripping party (Collection of William Bryant Logan)

Young men clambered along the fallen trunks and into the branches to strip the bark there. The children carried the bark away and stacked it to dry.

Once the rough outer bark had been pared away with knives and the tannin-rich inner bark ground as finely as possible in stone mills, the tanbark was ready for the tanner. But before the hides went into the tanning vat, the tanner had to prepare them to receive the fluids.

He'd first drop the fresh hide into water, let it soften, then "beam" it—drape it over an inclined log. He'd lean against the log to hold the hide in place and scrape the flesh side vigorously to remove as much of the fat and meat as possible.

Then he'd throw the hide into a vat of limewater, where it would stay until the hair began to loosen, then haul it out and wash it in water. Back to the beam the hide then went—hair side up, this time—and the tanner would carefully but firmly work the hair off the skin until the grain was revealed. Flipping the hide, he'd scrape the flesh side once more.

The effect of all this rough and heavy handling was to remove the hair and outer layer of skin, the epidermis, from outside and the fleshy layer from the inside. What was left was the dermis proper, a network of proteins composed almost entirely of collagen.

Before the tanning itself, however, the smelliest step was necessary. The tanner threw the hide into a bating vat, where it would steep in a brew of water and fermented chicken dung (dog or pigeon droppings might also be used). Two or three times during the next few days, the tanner stirred the hide around the vat, then drew it out to scrape it on the beam, then put it back again.

How this part of the process was invented is hard to guess, but it was indeed crucial. As it turns out, the ammonia in the fermented dung reacts with the lime left in the skin from its previous bath, extracting the lime and leaving the skin's pores completely clean and open.

Into the tanning vat went the hide. The tannins from the oak bark entered the voids left in the skin, creating chemical crosslinks with the collagen and creating a new substance, leather.

It often took as much as a year and many changes of tanning liquor before a large hide would be fully tanned and ready to finish, but the product was at last more or less waterproof and impervious to decay.

The tanner or his partner, the currier, would then finish the leather, beaming it again and then stretching it on a table to rub oils, fats, greases, and dyes into the surface in order to make the leather supple and to give it color.

Tanners and curriers had a reputation for tenacity well earned in the practice of their craft. Homer made a simile of their strength in the *Iliad*. Two armies fought over the body of Patroclus, here in Alexander Pope's translation:

> *As when a slaughtered bull's yet reeking hide,*
> *Strained with full force and tugged from side to side,*
> *The brawny curriers stretch; and labour o'er*
> *Th' extended surface, drunk with fat and gore.*

The tanner's work may have been rough and stinking, but he was a good person to have by you in a tough spot. One of the great tributes ever to the tanner is from the tales of Robin Hood. There is no one that Robin can't best in a fair fight, except for a man named Arthur Bland.

The two meet in Sherwood, each setting foot on the same log bridge at the same moment. Neither will step aside for the other. A fight ensues, each man wielding an oaken staff. Robin draws first blood, but Arthur not only cracks Robin's pate, he knocks him off the bridge.

As the outlaw climbs back for another round, he asks his opponent his name. In Howard Pyle's version, this is the reply:

> *I am a tanner, bold Arthur replied,*
> *In Nottingham long have I wrought;*
> *And if thou wilt come there, I vow and I swear,*
> *I will tan thy hide for nought.*

The two make up the quarrel, and Arthur becomes Robin's strong right-hand man.

Ink Forever

Permanence, dignity, strength, grace. These are the qualities that men and women sought in the oak. The same oak in the forest stood from generation to generation. The very form of the tree suggested patient power: It could hold a three-ton branch out sideways as far from the trunk as the trunk was tall. Collectors of older antique furniture often are collectors of oak furniture, not because people only built in oak, but because it is largely oak that has survived the furniture beetle. The other woods are long since heaps of powder.

Little wonder then that the ancients should search the oak for other ways in which its peculiar virtues could be employed. Acorns, wood, and bark were all important to them, but so were oak galls. The name, from the Latin *gala,* in fact means "bitter," so people must actually have tasted them. Yet in the bitterness, caused by a concentration of tannins, lay the secret of its usefulness. Crush the gall of the kermes oak, and you get a lovely scarlet dye, fitting for royal apparel. Crush the gall of the Aleppo oak (or of many other oaks for that matter) and mix them with iron, and you get a deep blue-black dye. Pliny knew of these dyes. He also noticed that an iron nail driven into oak wood would acquire a halo of black discoloration. Had he traveled in Africa, he might have seen a river that ran black, where a stream rich in dissolved iron met another stream rich in tannic peat.

Ink had been invented, probably independently, by both the Chinese and the Egyptians not long before Theophrastus had made his observations about galls. Both versions of the writing fluid used soot—derived from the burning of oil, resin, or tar—mixed with a vegetable gum or animal glue to help the color flow and to bind it to the page. These inks were serviceable but not long lasting. Because the tincture bound to the surface of the page, it was easily smudged or erased. In fact, it was not technically "ink" at all, since the word *ink* derives from the Italian *inchiostro,* the French *encre,* and ultimately Latin *incaustum,* meaning "burned in." The early inks did not "burn in," but lay on top of the page.

Sometime near the beginning of the Christian era, it occurred to someone that the blue-black dye made with iron and oak galls might make a better ink. The oldest surviving recipe comes from Martianus Capella's fifth-century *Encyclopedia of the Seven Free Arts.* Combining vitriol (an iron-sulfate solution that occurred naturally), crushed oak galls (for their gallo-tannic acid), water or wine, and gum arabic (a binder derived from acacia sap), the recipe was essentially the same as the one that would make iron gall ink the chief writing fluid of the Western world. This was a true ink, because the solution sunk quickly into the paper and there oxidized, turning black and binding to the very fabric of the page.

For five hundred years more, however, this indelible ink appears to have made little headway. It is a substance that cannot be made without healthy, far-reaching trade. The best vitriol was obtained from mines, as a solution that dripped from the rocks and was caught in barrels. It contained a number of minerals but was rich in iron sulfate. The Greeks called it *chacantum* (blood of copper) and the Romans named it *attramentum* (black-making). The most productive mines were in what is now Germany.

The gallo-tannic acid can be derived from many oak galls or even from oak or chestnut bark, but by far the highest concentrations are found in the gall of a scrub oak that grows abundantly in and near Aleppo in what is now Turkey. The tree is very unprepossessing for an oak—it grows but six feet high—

but its galls are powerful indeed. The tree took its species name from its use for ink making: *Quercus tinctoria*, the ink oak.

Gum arabic is the congealed sap that exudes from acacia trees native to the eastern Mediterranean and to Egypt. It is an ideal medium because it is soluble in water, flows well, and maintains the pigment particles in suspension. Though egg white was a commonly available binder in Europe, it was seldom used for gall ink.

Indeed, gall ink was rare in Europe until the Crusades, but beginning in the eleventh century—when the ingredients became readily available—it rapidly increased in popularity. It was far superior to soot ink—called "india ink." It flowed onto the page better, made a cleaner, sharper line, and, above all, once set it was virtually indelible.

Though it goes on only a pale gray, the gall ink oxidizes in the writing paper, producing a new ferric-tannic pigment— deep blue black in color—that binds to the page and is not water soluble.

Where clarity and permanence were wanted, oak ink was unsurpassable. Governments used it for their official documents. Among the first modern government contracts were those for the ink supply. The U.S. Constitution and the Declaration of Independence were both drafted with it. The German government used it continuously for three hundred years, ceasing only in 1974.

Architects used it for their drawings. The elevations that Thomas Jefferson drew for Monticello and for the University of Virginia were both in written gall ink.

Artists used it for their drawings. Leonardo da Vinci's note-books were all written with it. Bach wrote his music with it. The likes of Dürer, Rembrandt, and van Gogh drew with it.

Then too, every bourgeois housewife was supposed to be able to make it for her family. Recipes proliferated and people debated how best to prepare the galls. Some ground them, others boiled them, and still others let them mold. The moldy galls, it was said, made the best ink, since the mold concentrated the necessary acids.

Each thought that they were making a thing for the ages. Da Vinci's helicopter or van Gogh's trees or Bach's cantatas or Frau Van Vechten's husband's business correspondence would be preserved for posterity *sub specie aeternitatis*. And so for the next three hundred years, it was.

Then a funny thing happened. A librarian would open an old manuscript and a black dust would fall out. Within, the pages were eaten clean through, and the writing or the drawing was gone. Maps lost their boundaries, drawings their outlines, and illuminated manuscripts their texts. It did not matter whether the substrate was paper, parchment, or papyrus. The "permanent" markings were gone.

For a century now, librarians and conservators have been working intensively to find a way to preserve documents written with oak gall ink. The iron compounds, it seems, along with other metals, like the copper and the zinc included in the original vitriol, do not become inert but continue slowly to oxidize and react with the organic substrate of the papers, causing the degradation of cellulose.

It is a nightmare image to anyone who loves books: You

open to the page you seek and find that it has been destroyed not by fire or water or crumbling paper, but by the writing itself. To the writer, the artist, the calligrapher who pores over the work of his predecessors, it is like finding out that mother's milk is poison. The words, the written words themselves—the source of the music, the source of the meaning—are the source of destruction as well.

END OF THE AGE

FROM THE FIFTEENTH to the nineteen century, the power and reach of western and northern European cultures grew by powers of ten. From trading in the Baltic, hugging the Atlantic coast, and taking the dangerous leap across open water to Iceland, Greenland, and Vinland, the merchant adventurer began to cross the open sea. From journeys of a hundred miles, they went to voyages of one thousand miles, and then to travels of ten thousand miles, crisscrossing the globe.

Fed by the new trade, European wealth also grew by orders of magnitude. Trades that had never existed in the history of the earth grew and established themselves in the space of two or three years. New worlds were discovered almost weekly. News was really new, since a ship returning from around the Horn might be carrying cloth or people or art or plants or animals never before seen or described in Europe. Rice, potatoes, corn, sweet potatoes, tomatoes, peanuts, cocoa, peppers, sugar, spices, tea, coffee, cocaine, opium . . . once exotic or unknown produce became normal parts of European life. Medieval Englishmen never even heard of tea or coffee, but sitting in his coffeehouse in eighteenth-century London, Samuel Johnson could remark, "There is no illness so obstinate that it cannot be cured by the drinking of two hundred cups of strong tea."

The world for the first time experienced itself as a whole. Political and economic organization also grew by orders of magnitude, to gather and direct the resources needed for adventure. States and navies came into existence. By the end of the period, the merchants and their naval protectors had fully justified the boast inscribed on a seventeenth-century naval medal: *Nec meta mihi quae terminus orbi,* "Nothing stops me but the ends of the earth."

Sailing ships made all this possible, but in order to do so, they too had to change exponentially. Forty tons of cargo— what a good traditional Norse or Celtic ship could carry—was no longer nearly enough to repay the cost of a risky, four-year voyage. Four hundred tons was better, and better still was twelve or fifteen hundred. Furthermore, to cross the big waters, the oak timber in the old boats was far too small. A ship with planks three or four inches thick and a light frame not much thicker would break apart like matchsticks in the Roaring Forties south of the capes, where winds of one hundred miles per hour blew straight around the world without ever touching land and where the seas were often forty feet high.

In order to make ships exponentially bigger and tougher, shipbuilding was turned on its head. Up until the fourteenth century, the longships, the knars, the cogs, and most of the larger boats of Europe were built shell first. One plank was fitted and riveted to the next until the entire structure had been raised. Then a frame was inserted into the shell to improve its stiffness and to keep the rivets from working loose in rough seas.

Forest-grown oaks were best for these boats. They had grown in competition with other trees—in the Scandanavian world, they had had to outgrow evergreens to survive—so they tended to be straight and tall, with few branches until high up on the tree. Such trees contained good, straight-grained lengths for the long planks and strakes needed to form the shell-built boat.

The new ships were more like timber-framed houses than like earlier boats. They *began* with the frame—the skeleton— which was scarfed, lapped, and mortise-and-tenoned, and maybe bolted together. Only when the entire frame was raised was it covered with a shell. And the shell's planks were attached not to each other but directly to the frame.

Tall, forest-grown oaks were still needed to plank these ships, but huge oaks grown in the open were now the most important ingredient. The skeleton of each ship called for hundreds of structural pieces: ribs, knees, futtocks, floors, breasthooks, keels, keelsons, sternposts, and wing transoms. In an oceangoing ship, these timbers needed to have finished diameters of between ten and twenty-five inches, and they had to be found in the exact shapes required to make the curves of the frame. Only with such a stout skeleton and with well-caulked planks could ships sail the open ocean and around the capes.

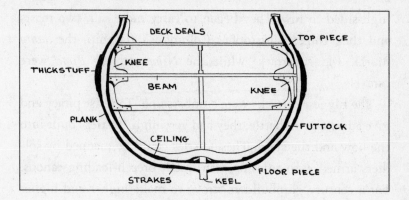

Frame construction of a 74-gun ship
(Nora H. Logan, after R. G. Albion, *Forests and Sea Power*)

Under this system, only the girth of oak trees limited the size of the ships that could be built.

More important to the merchant adventurers whose trade demanded such ships, the big boats had bottom. Indeed, the word *bottom* came into common use for the first time to describe the merchant ship's broad, deep holds. A merchant was, in fact, spoken of in terms of how many bottoms were in his fleet. From this word and usage came its application to human hinder parts, and also the common nineteenth-century saying describing a shallow man as "having no bottom." The bottom was where the wealth lay, and it also was the ballast that let the ship ride steady in rough seas.

In the fifteenth century, the new, skeleton-built ship bore different names in different parts of Europe—*nef, nau, carrack*—and it evolved rapidly in form and size. The caravel, a smaller ship than the carrack, was converted from shell to skeleton building at about the same time. All the new ships were too high-sided to row. They began to carry more sail, two masts and then three. Christopher Columbus's flagship, the *Santa Maria,* was a carrack, while the *Niña* and the *Pinta* were caravels.

The biggest carracks were cumbersome. To resist piracy and give advantage in battle, they had very high "castles" built into the bow and the stern. These castles could be manned by soldiers armed with rifles or with small, breech-loading canons. For a century, shipbuilders strove to make higher and higher castles, since then a ship's crew could shoot down at the enemy.

Unfortunately, high castles made the ships cranky and hard to turn. It also made it virtually impossible to keep them on a straight course. In this regard, they were the opposite of the sleek, fast Viking longships. Sir John Hawkins learned this to his cost, when in 1567 he fought the Spanish at San Juan de Ulloa in Mexico. Two of the British ships in his squadron, both small carracks, were able to maintain a tactical advantage, but the seven-hundred-ton *Jesus of Lubeck* was so unmaneuverable that the Spanish ran her down. She was lost with her entire crew.

A decade later, Hawkins had charge of Queen Elizabeth's navy. His predecessors under Henry VIII had already begun to cut away the towering castles and to place larger and more efficient cannon belowdecks, where they both were more potent and better balanced the boat. He probably knew too that the Dutch were experimenting with ships that had a very low profile to the water. He and his cousin, Sir Francis Drake, championed a whole new class of faster, more maneuverable ships for Her Majesty's navy. Drake's own *Golden Hind* was a good example. When the Spanish Armada approached the British coasts in 1588, the new British ships were instrumental in its defeat. The Spanish high-castled carracks were no match for them.

The British took the lead in shipbuilding and never relinquished it until just before the end of the era of sailing ships. The vulnerability of an island nation to invasion—a lesson learned well through the Viking years when the Norse both harried and conquered large parts of the British Isles—made the navy the first priority. The British navy's oak ships were dubbed "the wooden walls of England," and for the more than two centuries, from the destruction of the Spanish Armada to the end of the Napoleonic Wars, those walls were never breached.

Until the age of the nuclear-powered navy after World War II, there were no strategic instruments so powerful and so far-ranging as these skeleton-built oak ships. They could carry large amounts of cargo. They could spend months at sea out of sight of land. They could mount heavy cannon. Above all, unlike even the great battleships of the two world wars, they could be repaired, refurbished, and resupplied virtually anywhere in the world. Any forest could furnish wood to patch a hole or make a mast. A merchant voyage in the eighteenth and nineteenth centuries might last better than three years. Slowly but surely, these ships knit the globe together, bringing cultures that heretofore had been but shadowy legends into daily contact.

It is hard to comprehend the scope and depth of the change that overcame the maritime nations of Europe, from Portugal and Spain to the Low Countries and England. Unprecedented wealth appeared on their shores. To keep track of the wealth and to dispose it to advantage, paper currency became a potent force and stock markets were created. The old institutions creaked, groaned, and broke under the strain of so much wealth in so many hands. The wealth of kings was now but one among a number of power centers in the emerging nations, and the other holders of wealth resisted royal claims on their purses.

Kings could fall when they sought to exercise their royal prerogatives too freely. Early in the period, the English kings fell in love with big ships. Each wanted the best and biggest. James I had Phineas Pett build the *Prince Royal,* a ship of twelve hundred tons, almost a third larger than any other warship then in existence. It cost the unheard-of sum of twenty thousand pounds, including twelve hundred pounds just for the decorative carving.

To get the money, James invoked a primitive tax called "ship money." Since time immemorial, the English had followed a derivative of the Viking ship levy law: In time of war, any monarch could call on his coastal subjects to provide ships and men for the navy; if a county did not want to provide the ships and men, it could instead give money. In practice, this levy was now almost always collected in cash.

There was considerable grumbling both about the tax and about the size of the ship James built with it, but his son out-did him. Charles I wanted a ship even bigger than his father's, and the obliging Phineas Pett laid out for him the *Sovereign of the Seas,* the first three-decker ever conceived. It would have three covered decks, each loaded with cannon; no other ship had had more than two. It took twenty-eight oxen and four horses to drag the immense keelson (the interior part of the keel) from the Weald of Kent down to the sea. Furthermore, the *Sovereign* was even more decorated than James's ship, and its tonnage was 25 percent larger.

To pay for the *Sovereign* and for other new ships, Charles again invoked the ship money tax. But there were three problems: First, ship money had never been collected except in time of war, and England was then at peace. Second, ship money had only been collected from coastal towns and counties, but Charles insisted on collecting it from inland regions as well. Third, he sought to make the payment a permanent institution. Essentially, he was instituting universal taxation, but he did so without the consent of Parliament, and opposition was intense. Some refused to pay it, others refused to enforce it, and though the courts sided with the king, the outcome was civil war.

When war broke out, the ill-paid naval officers and crews went over as a body to the Parliamentarians. Charles I was executed, his two sons, Charles and James, fled to France, and England embarked on two decades of parliamentary rule under the leadership of Oliver Cromwell.

Twice the younger Charles tried to return to his throne. The first time, he never succeeded in making landfall in England, but the second, he led thirteen thousand men south from Scotland, across the English border, only to be soundly defeated by the Parliamentarians at Worcester. By his own account—as reported by Samuel Pepys in *King Charles Preserved*—he escaped thanks to an oak tree in whose branches he hid. "We carried up with us," Charles reported, "some victuals for the whole day, viz. bread, cheese, small beer, and nothing else, and got up into a great oak, that had been lopt some three or four years before, and being grown out again, very bushy and thick, could not be seen through, and here we staid all day." Thereafter, he escaped to France once more.

After Cromwell's death, there being no institutions to choose a successor, and to prevent anarchy, Parliament invited Charles to return to England as Charles II, but only under certain conditions: that he would grant amnesty to all except the men who had actually killed his father, and that all his acts would have to be ratified by Parliament before they were considered valid. In other words, he would be subject to law.

In the Parliament of 1677, the king, represented by his clerk of the admiralty, Samuel Pepys, expressed his wish to build thirty new warships. England was at peace, though the king had recently promulgated a disastrous war with the Dutch, relations with Parliament were strained, and Charles half-expected that his request would be denied. But Pepys persuasively showed the need for the new ships, specifically comparing the strength of the English navy with the navies of its principal rivals, the French and the Dutch. It was the first time in modern history that a minister had invoked an arms race in order to get his point across, but the tactic worked. Parliament voted the money for the ships.

Still, many of the representatives believed that the ships were Parliament's or the people's, not the king's. One wrote, "What we give, we give not to the king, but for our own defence." And indeed, the bill as passed mandated that all the monies for ships be kept separate from any other monies given the king, and that the ships be completed on schedule. There were penalties for lateness to be levied against contractors and even against the king himself. A general tax on all real property would raise the funds.

This was the birth of a modern state. Above all else, the state promised protection to its citizens. No one—not king, not noble, not churchman, not pirate—was to be above the government. The state itself, with its permanent corps of sailors and soldiers loyal not to individuals but to the institution itself, would guarantee its sovereignty. In return, the citizens would be free to make their own lives as prosperous as possible and would pay the state taxes to enable it to protect them.

In England, the system remained a hybrid, but more and more, the king was subordinate to Parliament's will. Parliament had essentially given Charles II exactly what his father, Charles I, had peremptorily taken. But the son, among whose stated objectives was "not to go on my travels again," took Parliament's largesse kindly, even with the strings attached. His minister Pepys, who would become the prototype of a valuable bureaucrat, made sure that the ships would be well manned by getting His Majesty's consent for further reforms.

It had been customary in the past for captains to use state ships for their own purposes—chiefly to carry cargo—when not on active duty, in order to help supplement their meager pay. Pepys had the pay of all officers and sailors immediately more than doubled, and he strengthened the prize system. Once, the Crown had taken up to half the value of a captured enemy ship. Now, the officers and men who took the prize were to keep the entire value, each receiving a fixed share. Officers might grow wealthy on their prize money, but even an ordinary seaman might put aside enough to buy a house or a farm on shore.

Pepys was himself a symbol of the change coming over England. At once a servant of the Crown and member of the House of Commons, he extolled the king while serving the interests of the nation.

Even the king embodied this mixture: In part, he was a courageous national leader, whose personal knowledge of and interest in warships assured that good ones would be built. No decorations for Charles II, no fancy royal names, no single great ship. Instead, he sought thirty ships of the line, low-slung but strong, with only broadside cannon, meant not for single action, but for battles between fleets. On the other hand, in his personal life, he supported a string of mistresses, so that part of Parliament's strictures on the ship tax was meant to keep it away from the king's ladies. Furthermore, nostalgic for absolute power, he conspired ineffectually with the French to restore Catholicism to England and, with it, the divine right of kings.

In 1677, fully half the revenue brought in to the treasury went to build the thirty ships. More money went for the decent pay of the officers and crews, who would now be paid in peace- and wartime both. Permanent shipbuilders were hired, who became experts in making very large ships dedicated to warfare. The ships protected not just royal interests. They also convoyed merchantmen and created a permanent threat to those who would harass trade on the high seas. They controlled and eventually eliminated private, as opposed to state-sponsored, warfare.

This emerging state supported every kind of private enterprise—including some notable speculation in shipbuilding and increasing bureaucratic corruption—but it resolutely limited the private business of violence. Piracy and private armies were suppressed at home and abroad. It was indeed possible to be a "privateer" at sea, but only with a specific license from the state. An unlicensed pirate could be hanged on capture, without further ado. The state was now the sole supplier of violence and the sole guarantor of liberty and the rule of law.

The most successful country to make this transition, England, was also the winner in three centuries of European confrontation at sea. But even in absolutist France, the change from king's ships to national navies began in this era. Louis XIV's great minister, Jean-Baptiste Colbert, capably created a set of policies of forestland preservation and shipbuilding that were the envy of Europe, including England.

English nobles owned nine-tenths of the land from which the big, open-grown oak would come, and in the House of Lords they defeated every measure meant to restrict the use or control the price of oak. Colbert, on the other hand, promulgated a forest law that marked and preserved French oak for French ships. No timber could be cut within fifteen leagues of the sea or six leagues of a navigable river, unless a written application was submitted and a waiting period of six months observed. He enforced the policy rigorously and prevented the formation of trusts that could force up the price of oak. Furthermore, he rewarded fine shipbuilding and oversaw the development of the seventy-four-gun ship, the size and draft that would become the standard for large warships through the rest of the age of sail.

The Dutch, among others, increasingly fell behind in the

maritime competition, because they did not make this transition. While England and France were building professional navies, the Dutch continued to count on their merchantmen to be converted into warships at need, supplementing a few, generally smaller, dedicated warships.

For two and a half centuries, the nations of Europe clawed each other for the prize of naval supremacy. The English were always in the center of the fight. In the seventeenth century, their opponents were most often the Dutch. In the three Dutch wars, little was decided dramatically at sea, though some battles went on for days. The line of battle—a tactical philosophy under which head-to-tail lines of the largest warships ran alongside one another, battering away—was first conceived during these wars. At the start, the Dutch had three merchant ships to every one of the English. At the end, the situation was reversed.

At the beginning of the eighteenth century, a third of the world's naval power was English, another third was split by the French and the Dutch, and the rest of the world comprised the last third. The French and the fading Spanish were England's combatants in this century, and at Malaga in 1759, it looked as though the French might prevail. They soundly defeated Admiral Byng's squadron and captured Minorca from the English. Admiral Byng was court-martialed and shot.

But from the Battle of Quiberon Bay in 1759—before the French Revolution—until the final exile of Napoleon Bonaparte in 1815, the British navy dominated the French and their allies. Five French ships of the line were taken at Quiberon Bay by Admiral Sir Edward Hawke. Two decades later, at Cape St. Vincent, Admiral Rodney fought the Spanish allied with the

French, destroying one ship and capturing six. At the Battle of
the Saintes near Dominica in the West Indies three years later,
Rodney sank one French ship of the line and took five more.
On the Glorious First of June, in 1794, Lord Howe took two
French eighty-gun ships and four seventy-fours. A fifth
seventy-four he sank. Lord Nelson, at the Battle of the Nile in
1798, destroyed eleven of the thirteen French warships arrayed
against him. Finally, at Trafalgar, Nelson put an effective end
to French naval power in an overwhelming action through
which the French lost eighteen ships. Nelson himself died in
the battle.

By the start of the nineteenth century, oak ships had delivered Europe from the medieval into the modern world. Every
coast with any resources at all was the partner or the colony of
a European (or American) power. Individual liberty was a principle at home—albeit a very contentious one—but liberty was
absent abroad. Oak ships had made the world a single, but
unequal, community.

Into the Woods

The great, skeleton-built ships began in the woods,
where imagination had to find the ships' actual materials. Whether it was to be a sloop with a crew of twelve or a ship
of the line with a crew of five hundred, the boat would be at
least 90 percent oak, including all of the structural timber.
John Evelyn—a friend both of Pepys and of Charles II, whose
1664 *Sylva* was forestry's founding book—summarized why

oak was preferred: "tough, bending well, strong and not too heavy, nor easily admitting water." He might also have added that oak, alone among the trees of the temperate forest, regularly holds out immensely long branches at ninety-degree angles to the main trunk. Its branch crotches are wide and strong, creating just the right shapes for compass timber.

The forester was as tuned to the health and shapes of trees as a physician is to the human body. He was looking for two things: big bent or crotched pieces of compass timber for the frames of the ship and long, straight timber to be sawn into thickstuff and planking. But neither the right size nor the right shape would matter if the tree were unsound.

Into the woods he went: in Sussex or in the Welsh marches; in Normandy, Denmark, or Germany; in Connecticut, New Hampshire, or South Carolina. Jays and squirrels complained of his presence. His boots cracked twigs and sank in the deep duff. Where the trees grew in close mixed forests, the trunks would tend to be straight, with few branches until high in the crown. Where the trees grew in the open, they would have wide-reaching branches from bottom to top. Those were givens. Everything else was a matter for the forester's judgment.

First, he'd look at the lay of the land. Was there a lot of projecting rock? Maybe the soils were thin. In that case, he'd look for windthrown trees to see how large they were. A tree knocked down by the wind usually had some decay at the base or in the roots. Looking around the forest, he'd judge how many trees were that size or smaller, figuring that the larger ones likely had at least some decay in their stems. The larger trees were unlikely to make good ship timber.

Were there trees that indicated poor soils? In American forests, black oak *(Quercus velutina)* grows where the soils are poor. Nobody would ever build a ship with black oak. Unlike the trees of the white oak group, its heartwood is porous, but white oaks growing in that vicinity were likely to be more stressed, less sound.

Maybe in some parts of the forest, the land leveled and the soils deepened, maybe at the base of a ravine or along the outwash of a brook. He'd pay special attention to the trees in that area. They'd likely remain sound longer than hillside trees.

He'd look how the land faced the wind and the weather, whether the trees were sheltered or exposed. If the trees were high on a slope, exposed to sudden freezing and thawing, or to heavy winds, they might have star shakes or cup shakes inside.

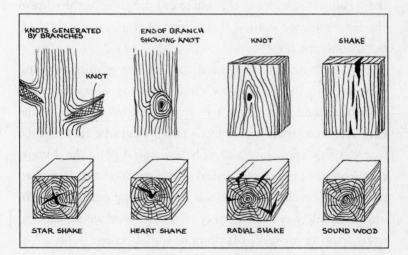

Some defects that ruin a ship's timber (Nora H. Logan, after
J. C. S. Brough, *Timber for Woodwork*)

A star shake often begins with frost cracks formed when the bark of the young tree is exposed to winter sun; the frost makes a star-shaped pattern of fissures inside the trunk. A cup shake runs right around one or more of the annual rings, shaking it loose from the rest of the trunk. Either shake rendered the timber useless. If it were for compass pieces, the shakes would make it weak and unable to hold its place in the frame. If it were for plank, the plank would break into pieces when it was sawn out.

All oak tends to flare at the base, knew the forester, but a too-dramatic flare suggested decay. How far up did this rot extend, and would it make the whole tree useless? It was hard to tell. Even were the wood above relatively sound, it might develop a heart shake, from the impaired ability of the rotten center of the tree to hold moisture. A heart shake is like a star shake, only it is wider at the center and narrower at the edge of the circumference. Still, if it extended, it would make the timber fit only for firewood.

He'd look too for cat faces, if the tree was promising plank material. Unclosed or barely closed wounds where branches have been shed often show a roundish mottled pattern on the bark that resembles a cat's face. There would be flaws in the plank at that spot. Likewise, if he saw spiral grain, he'd know that that tree's tissue was loaded under tension, and that when it was cut, it would likely release the energy of that bending in unpredictable ways. At the very least, the wood would want to warp, making for leaky planking or sprung knees.

When he'd chosen the trees, he'd stand by while they were cut. He waited for the smell of each. A sharp sweet tannic odor meant the butt was healthy and so, likely, was the tree. A musty or rotten odor might mean the wood was already decaying, even if it appeared sound. He'd look for any patterns of discoloration or for the presence of the black threads of fungi. He'd sound the log up and down with a hammer, listening and feeling for hollowness. The great French forester Garaud listed twenty-seven common defects in standing timber and thirty-eight more to be looked for once the tree was felled. Just as in the days before limited liability corporations, a trading firm might be rich one day and broke the next, so a tree that looked beautiful and stately in the forest might turn out to be totally rotten. John Evelyn was not exaggerating when he remarked, "The tree is a merchant adventurer. You shall never know what he is worth until he is dead."

Even within sound wood, the forester might begin to make distinctions. If he shaved off a chip with his knife and it held together in one piece, twisting into a spiral, he knew he had lean oak with long fibers, hard to wrack out of shape and good for framing. If on the other hand, the chips were short, flaky, and easily broken, they indicated fat oak with short fibers but likely with excellent water-repelling properties. This wood would be better for planking.

The fellers' work called for sound judgment and for care in equal measure. In recent years, machines have been developed that grab trees by their trunks, jerk them from the ground, strip off the side branches, and deliver the valuable trunk wood to the log truck. What a savings in labor it is, one the wood-cutter would have admired. But he would have shouted in dismay over the waste: trunks cut carelessly high, beautiful bends at the root flare or at the branch crotches destroyed, tops left lying in heaps of refuse.

This was the work, he might have said, of a poor jobber, not a feller, not a woodsman. The craftsman went to his work with a rhyme in his head, such as the following: "Sell bark to the tanner ere timber ye fell, / Cut low to the ground, else do ye not well, / In breaking save crooked, for mill and for ships, / And, ever, in hewing save carpenter's chips."

Nothing went to waste. He looked at the shipwright's molds, then found the same shapes in the trees. If the ship wood was in trunk and boughs, he would cut away the root flares to leave a smooth bole. Then, he'd make a face cut in the side toward which he wanted the tree to fall, making the cut as low on the bole as he could. This was usually stoop work. Then he'd make a back cut behind it, a little higher, but as little as possible. Early on, this was done with the same ax as cut the face. Later, a saw would be used to open the back. The sawyers for this cut—as many as four on a big tree—might work on their knees, drawing the big two-handed saw back and forth with ropes attached to each handle until the tree went over.

The feller was a practical physicist. In a wood full of trees and undergrowth, he had to make his tree fall so that it did not hang up in any other tree and so that when it fell it did no damage to the beautifully shaped crotches in the branch wood. Otherwise, a sternpost or a half dozen first futtocks or a breast-hook—crucial and hard-to-find pieces on a sailing ship—might be spoiled.

He also had to prevent it from splitting in the bole, a hard job if the tree was a leaner or showed any defect at the base. If he suspected the trunk was going to split, he might reverse the usual procedure. He'd first saw in underneath the lean, halfway through the bole. Then he'd attack the root buttress on the uphill side with an ax, cutting out chips until the tree slid off its base in the direction of the lean.

If a "claw"—the angle from the base of the trunk into the root flare—was to be preserved for a knee or another curved piece of ship's timber, he'd have to saw out all around it, then make his face and back cuts so as to leave this precious shape whole and safely in the air.

Much is revealed in the moment when a tree at last begins to fall. If the feller has done his work well, it will fall more or less in the direction it should. But if the tree is not sound within, it may split, twist about its vertical axis, and fall in an unexpected direction. This barber-chairing, as it is still called, could kill men and ruin timber. Once the notch and the back cut are made, the feller watches the tree begin to fall. Occasionally, it does not seem to fall at all. There is a moment of fascination. "Now my mind tells me the cuts are made and the tree is falling," the feller thinks. "Why isn't it falling?"

If he is smart, a moment later he runs just as fast as he can to one side or the other. For the only position from which a falling tree does not appear to fall is directly in the line that it is falling.

When the tree was down, he took off branch wood and selected the shapes he wanted from the shapes he'd found, laying molds over the stems and branches. Others might go in to bark the tree, pulling off everything that could be used for tanning. Children and women would come in behind, carrying off the "chips"—the waste wood—to become firewood or charcoal.

Often, the foresters and fellers got the chips and the bark as their perk, and these were of substantial value.

From the felling ax, the woodcutter switched to the broad ax and began to rough out the big shaped timber for the ship's frame. He cut notches through the trunk, figuring how broad a piece he wanted. He repeated this same cut through the whole piece and to the very same depth. He worked crossways down the log, cutting away the notches, until he was left with a roughly smooth surface. Then with a cant hook, he'd flip the piece and do just the same work on the other side. By the time he'd finished, he'd performed what almost seemed a sleight of hand: What began the afternoon as a piece of tree ended it as the rib of a ship.

It was backbreaking work and subject to cruel disappointment. The crosscuts might reveal a black and rotten heart. The old sayings about men and women, that their hearts were black or rotten, come from this disappointment and exhaustion. In the midst of the hewing, a small cup shake might widen and spread along the whole length of the piece, again making a piece of wood weighing better than two tons into lumber fit for nothing but kindling. When we say a thing is "shaky" or that we are feeling "shaky," this is where the saying comes from: the unsound cracks hidden in the wood that make the whole piece weak and unstable. Hours of labor might yield not one sound piece.

The costliest part of the whole endeavor was getting the timber from the forest to the yard. A twenty-foot oak log of roughly twenty-four inches in diameter weighed better than a

The felling and measuring of straight, forest-grown oaks, suitable for a ship's planking (from Brian Latham, *Timber: A Historical Survey of Its Development and Distribution*, author's collection)

ton and a half. (The same log in the American live oak that formed the ribs for the USS *Constitution* would have weighed more than two tons.) The butt of the log would be "tushed" to the harnesses of oxen and horses, who'd haul the wood out to a beach. Twenty miles was the maximum distance for this haul, and even at that it was a matter of a couple hard days' work. This is the origin of the phrase "in it for the long haul," which means a person will stick to a task no matter how unpleasant.

At the beach, you would think the logs could be pushed into the water and rafted together. Unfortunately for the hauler, however, oak is just about as heavy as water, particularly when it is green. To get the timber to the yard by sea, the oak pieces had to be either rafted together with lighter pine and fir or swayed up and aboard large stolid hoys, barges that could carry them along coastal waters. Sometimes, pieces might be dragged up onto a wagon and transported by road to the yard, but conditions had to be right. New Hampshire Yankees could skid their wood on sleds in the frozen winter, but in coastal Europe, the teamsters might have to wait a year or more for the road to dry out enough to let the heavy loads pass. The cost of transport represented the greater part of the wood's price.

The Crafts Assembled

> *Take it all in all, a Ship of the Line is the most honourable thing that man, as a gregarious animal, has ever produced. . . . Into that he has put as much of his human patience, common sense, forethought, experimental philosophy, self-control,*

habits of order and obedience, thoroughly wrought handiwork, defiance of brute elements, careless courage, careful patriotism, and calm expectation of the judgment of God, as can well be put into a space of 300 feet long by 80 broad. And I am thankful to live in an age when I could see this thing so done.

—JOHN RUSKIN,
The Harbours of England (1856)

So wrote John Ruskin in his text for an edition of J. M. W. Turner's contemporary pictures of English ports. Here, he was looking at Turner's sublime images of the ships, lit dramatically and in ragged weather. They seem both insubstantial and larger than life. One wonders what Ruskin would have written if instead he had been standing in a shipyard.

There, the hard thing to believe is that the smooth and beautiful curves of the ships standing on the stocks about to slip into the water actually came from the irregular pile of logs and sticks—long and short, straight and crooked, big and little, some with bark, some with branches, some rotten or split, some blue at the butt, some brown, some yellow—stacked, jumbled, and dropped not one hundred yards away up the beach.

In between the upstanding ship and the massed wood, there are men hewing with axes, smoothing with adzes; pairs sawing baulks with pits saws; men measuring and cutting and measuring again; men sliding half-finished timbers along a corduroy road of trunks; men turning dowels and cutting trenails; men riggng and lifting wood with block-and-tackle and cranes; men scarfing twenty-five-inch-squared baulks into long keels; men boiling pitch and mixing it with oakum to make a mess of smelly black caulking; men hammering out rivets on anvils; men snugging on planks and hammering them in against ribs or on floor pieces. Someone calls from a half-raised ship skeleton, and two dozen men stop their work to come help crane the next frame onto the keel. A bell rings, and every man shambles, scrambles, ambles, or limps over to a shed for a ration of grog. The women come sweeping into the yard to carry away the waste wood for their home hearths.

Ruskin's wonder might indeed have increased had he seen the shipyard. For who can believe that out of such a rough pile,

A seventeenth-century shipyard (Ake Ralamb, *Skeps Byggerij eller Adelig Otnings*)

through such a bedlam of hacking, sawing, stewing, and lifting, comes a thing so beautiful and so capable as a great ship? Yet there they all stand together, and nobody has ever waved a wand to make the ship come to be.

Until the middle of the nineteenth century, the shipyards of Europe and America represented the largest industry in the world. In those yards was formed the division between thinking and making that is the hallmark of modern industrial production. The designers planned, the managers arranged, and the workers made what they were told. It was necessary to make skeleton-built ships in this way. There were hundreds or even thousands of pieces to each ship's skeleton frame. Each had to be cut to dimensions that would allow it to snug against its neighbors, and all had to create a smooth line. If they did not fit exactly, the boat would leak.

Yet through most of the age, the thinkers still made and the makers still thought. Feedback between thinking and making—such as was necessary to craft production—continued to some degree in the organized work of the yard. The designers whittled out their models; the plankers adjusted the oak to the actual curve of the frame.

The shipyard was not yet quite an assembly line, but rather an assembly of the crafts. There had to be a wright to carve a small model and to lay out the dimensions, a forester to find and select the right timber, an axman and a sawyer to cut it properly, teamsters to tush up the log and take it to the water, hoymen to load and transport it to the yard. In the shipyard, generally located on the firm sloping beach of a "hard," a tidal creek, there had to be pit sawyers to cut the planks and thickstuff; axmen and adze men to shape the frame pieces, beams, knees, and masts; riggers to set the keel on the stocks and to raise up the frame piece by piece; and joiners to make the trenails and to snuggle and fay, to join snugly and securely, each piece to its mate.

A skeleton-built ship was the opposite of a shell-built ship, not only in means but in the materials required. A Viking ship was 90 percent planks and only 10 percent frame timber. The carrack, the frigate, the sloop, the brig, the ship of the line—all were 75 percent frame timber, 15 percent thickstuff to stabilize the frame, and 10 percent ordinary planking to cover the sides and the decks. The average ship of the line required three thousand loads of oak, or about sixty acres of century-old trees. It took at least six months and sometimes up to three years to build.

First, a shaping imagination was needed. The great shipwrights, heirs to the tradition of the stemsmiths, knew that a sharper profile would give a faster ship, but that she would pitch and roll more, and not keep an even keel in battle. A wider profile would hold more cargo or crew and guns, but if too wide it might wallow and overpower the rudder, making the ship cranky and hard to steer. They looked for a medium between these, minimizing faults and maximizing speed and stability. Usually, they formed their ideas by looking at the work of their own teachers or at other boats they admired, sometimes even at boats captured from the enemy. They had not only to make the requisite shape of shell, but to render this planking stable with a strong, economical frame.

For most of the age of sail, the ships were too complex and too costly to be made by rules of thumb. The wright would carve a small half-model of the ship to be constructed, showing the lines of the ship from stem to stern. Owners or clients might argue and dispute over this model, but once it was accepted, it was the basis on which the ship would be built. It took eighteen tons of paper blueprints to specify exactly how the World War II battleship the USS *Missouri* was to be built, but the basis of a sailing ship was a small model easily held in two hands.

The wright delivered this model to the loftsman, who scaled up the model to full size, sketching out each piece of the ship on thin wood on the floor of a large room. These molds, templates of the shapes needed, would sometimes be given to the builders in the yard, who would seek the appropriate pieces on-site. Whenever possible, the timber-getters would take the templates with them into the field, to help in selecting the right pieces.

A mistake in size or shape was no small matter. The ships of the line not only took three to four thousand oaks—the American frigate *Constitution* took a mere fifteen hundred—but they required pieces of huge size and specific shape. Oak was never bent or constructed to make all the pieces of the up to seventy curved ribs, or the stem or sternposts, or the breasthooks, or the more than forty knees that stabilized the whole structure. Trees had to be found that had grown straight grained to the required sizes and shapes, their natural branching patterns approximating the shapes needed. A tall oak that split into two leaders near the top might make a sternpost. An oak that divided low on the trunk, grooving in a V-shape, might become a wing transom. The curved boles of wind-shaped hedgerow oaks might be turned into futtocks (the "foot oaks," or lower pieces of the ribs) or floor pieces.

On average, to be useful for a big ship, an oak had to be seventy-five to eighty feet long and yield timber more than twenty inches on a side. The sternpost of a ship of the line had to be forty feet long and twenty-eight inches thick. The seventy ribs each required at least four long thick curved pieces of oak. Each knee had to be bent like the knee of a seated human, but it had to be twelve to fifteen inches in diameter.

Even with models and templates, the ship that was finally launched might not behave as intended. The American whaler *Trident* was so lopsided it had to carry 150 extra barrels of oil on one side in order to sail straight. The huge Swedish *Wasa* was so poorly ballasted and balanced that she sank on her maiden voyage, shortly after setting sail, when a squall healed her over, allowing water to rush into open gun ports.

Almost all shipyards were located on what the English called a "hard." This was a clean, broad, sloping beach leading down to a tidal estuary with good deep water offshore. While the wood was gathering in, the first task was to make a slightly sloping base for the new ship to rest on. The whole enterprise depended on properly making this slip.

The word *slip* seems a slight one to describe a two-thousand-ton hulk sliding from the beach into the water. Imagine what it would be like to push a forty-room mansion into the sea, and

Ship timber found in the natural curve
of oaks (Nora H. Logan)

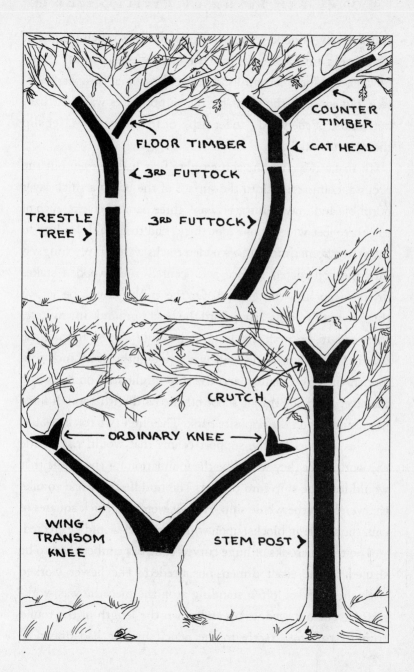

FLOOR TIMBER

COUNTER TIMBER

CAT HEAD

3RD FUTTOCK

TRESTLE TREE >

3RD FUTTOCK >

CRUTCH

ORDINARY KNEE

WING TRANSOM KNEE

STEM POST >

you will have a good comparison, for about as much weight of wood went into a ship as into a house. We seldom say "slip" now, except for a certain kind of accident on a "slippery" surface, but more to the point for the shipbuilders was the Elizabethan use of the word, "to let slip," or to release, as in "let slip the dogs of war."

To make the slip, the shipwrights first had to even out the ground, compacting into the surface of the beach a thick layer of rubble and gravel. Into this base, three parallel tracks, each of oak three feet wide by one foot thick, had to be laid, leading to the water. Sometimes these wooden tracks were simply snugged into the compacted surface and secured with wooden stakes driven deep into the ground. Or the tracks might be set on three rows of pilings driven into the hard's ground. In any case, they had to be perfectly level and evenly sloped toward the water. The recommended fall was five-eighths of an inch to a foot. A little too shallow and the ship might stall on the ways; too steep, and she might shoot off the hard with enough force to ground her on the opposite bank. The outer two tracks would serve as anchors for the support beams that would rise beside the work; later they would be the foundation for the cradle that would lead the ship into the sea. The middle track had to take the weight of the whole ship. Atop it were laid thick squares of oak, the splitting blocks, upon which the keel would be erected.

The rough crooks of huge curved compass timbers had to be shaped to the exact dimensions needed. The hewer worked with the broadax, often standing atop the piece he was working. He cut out wedges up and down the length of the round trunks, working exactly to the pencil line he had drawn to mark the squared dimensions of the piece. Then he'd cut through these wedges from the sides, splitting them out, until at least the faces that needed to fit in the frame were squared. The work was exacting and done exactly. "Hew to the line, and let the chips fall where they may" is said of any man who sticks to his task through difficulties, but it first refered to the axman in the shipyard.

The pit sawyers meantime cut straight logs into planks for the ship's sides. Either working in a pit, with the topman on the ground and the bottom sawyer below, or by raising the log onto trestles and working from trestle top to ground, the pair worked as a team. The topman aimed the saw and guided its progress, the bottom man pulled the blade through the wood. Sawyers were always hired in pairs and often spent their whole working lives together. They were not sawing out the three-quarter-inch pine boards with which we are familiar. They were sawing forty-foot-long strakes, some of which were six or seven inches thick.

Like most of the ship makers, they had learned their jobs from others. They had sayings to prompt them. "Strip when you're cold, and you'll live to grow old" reminded the sawyers not to get their clothes sodden with sweat and thus to avoid catching a chill when they rested. "Hard knots and empty pint pots are two bad things for sawyers" reminded them to watch out for knots in the wood, which might break teeth off the saw blade, a fate as evil as running short of beer.

Nothing went to waste. A stoutly defended perk was that three times a day the workers' women could come to take home the chips that had been cut, for firewood or charcoal making.

The sawdust went into blue-smoking fires over which tallow for graving or pitch for caulking was heated.

The dubbers followed both hewers and sawyers, finishing the compass timbers and the planks with adzes. Though the adze in many hands is a fairly rough tool, a good dubber could make an almost mirror-smooth surface where it counted—where two wood surfaces had to be fayed, or fit, quite smoothly. He didn't waste his effort where wood would not be touching wood, however. When you look at surviving tall ships, such as the USS *Constitution* (now afloat in Boston, Massachusetts) or Lord Nelson's HMS *Victory* (now in dock in Portsmouth, England), you see huge crooks of wood in the breasthooks and the sternposts, whose inside faces still show the curve of the branch, a strip of bark, and the collar where a smaller branch came out.

Meanwhile, riggers had set up sheerlegs, derricks, and jib cranes where the stem and the stern would start to rise. These cranes were the same as those built thousands of years before in the Mediterranean. At their top extremities they held systems of block and tackle so that men or teams of oxen could lift and settle heavy timbers on the keel.

The shipwrights supervised the cutting of scarfs on the huge lengths of keel wood, and watched as the pieces were fit in place atop the splitting blocks. The scarfs were joined with copper through-bolts, some more than three feet long. (The copper bolts for the *Constitution*'s keel were made in Paul Revere's foundry.) To hold the keel in place, thick trenails (oak or black locust pegs) were wedged diagonally beside them on the splitting blocks.

From this point forward, every man in the yard was alert for the call "Frame ho!" At the word, they would stop their appointed task and go to help with the raising, swaying, and fitting of frames. First came the stem and the sterns; then came the centermost, or main, frame. Then frames were filled in fore and aft. Each frame was assembled on the ground, then raised into position, fayed into a pre-cut scarf, and bolted through with copper bolts or trenails.

Whole towns were dedicated to the production of trenails. (For the Portsmouth yard in England, for example, the village of Owlesbury, near Winchester, supplied almost all the trenails.) Trenails, whether oak or locust, had to be well seasoned so that they would not shrink. Then, when the green planking seasoned and shrank around them a tight fit would result.

When the framing was yet incomplete, plankers would mount the rising scaffold inside and out. Working in pairs, like sawyers, they would coax and bang each plank into position until it was perfectly fair with its neighbors. If necessary, they would adze off a little more here or there, but mainly they strained, hammered, wedged, and bent. When the piece was fair, they'd yell out "wood on wood!" to let their partner know the pieces had fit snugly.

Frame, plank, decks, and the ship took shape. All the workers took grog three times a day, and at those same times, day after day, the women came to collect the chips.

The designer's role became more important as maritime competition increased. More than half of all the wooden ships ever built were constructed in the first three-quarters of the nineteenth century, the last years of the age of sail. The finest incorporated new ideas to increase speed and stability. For the

USS *Constitution,* completed in 1797 in Boston, Joshua Humphreys and Josiah Fox had added two bright ideas. First, some of the decking was cut extra thick with matching notches, one facing forward and one facing aft. These deck planks fit together like a jigsaw puzzle. The result was a fore-and-aft section on each deck that was extra stiff and extra hard to move, making the ship tighter and swifter.

More cumbersome to install were the huge diagonal riders that they specified to run from the keelson—the piece that fit atop the keel in the ship's inner skin—diagonally up and forward until they tied into the rising frame of the ship at large standing knees. Only a few ships in Europe had yet tried this idea when it was installed on the *Constitution.* The point was to prevent the ship from hogging, riding up at its middle and so slowing its way. Their effectiveness was proven in the later years of the ship's long life. When the diagonals were removed from the *Constitution* during the early twentieth century, she had a hog of 16 percent. When the riders were reinstalled, the hog was immediately reduced to 4 percent.

When a ship was ready to launch, the slip's two lateral tracks suddenly became important. Using these as a base, the ship-wrights built a cradle, or bilge way, on either side of the finished hull. This cradle was well lubricated with tallow. It would guide the ship into the water and prevent it from keeling over, which would ruin two years' work in a moment. Again, the angle was crucial. Too shallow and the ship might hang up. Too steep and it might race out of control across the river. The worst result would be for the ship to stop half in and half out of the water, since the forces of sudden buoyancy on only half the ship might break her in two.

The shipwright's last task was to scarf and bolt the false keel beneath the keel. This sacrificial piece was also called the worm keel, since it was meant to suffer damage—whether from boring worms or running aground—while protecting the true keel, upon whose integrity the structure of the ship depended.

One by one, they knocked out the splitting blocks and let the ship rest on its false keel, straight on its track into the water. They installed a screw mechanism facing the stems, well anchored to the ground. Often, too, they rigged blocks and tackles from the existing cranes to help encourage the ship down the ways.

When it was time to launch the *Constitution* in 1797, the naval constructor, Col. George Claghorn, had every dignitary present from President John Adams down to the customs house clerks. He knew that her sister ship, the USS *United States,* had gone too fast down the ways in Philadelphia, had piled up on the opposite shore of the river, and had had to be repaired before ever she put to sea. Perhaps this inclined him to be too cautious, for with thousands in attendance, the *Constitution* scarcely budged on the ways. She went about twenty-seven feet, then hung up, a portion of the track apparently having settled.

Everyone went home, while the opponents of naval expenditures began to hoot and holler. That afternoon, with no one present, Claghorn tried again. This time she got to within a few feet of the water, but again hung up. Imagine the frustration of men who had put two years of their lives into this majestic structure, only to find it refusing to enter the sea.

But he'd lost his tide. Claghorn had to wait a month for the next water high enough to float her. Finally, in mid-October, when the last red and yellow leaves were blowing off the maple trees in Boston and with no one but the builders present, the *Constitution* journeyed the last few feet down the ways and floated on her own bottom.

Constitution

After Nelson's crushing defeat of the French navy at the Battle of Trafalgar in 1805, Great Britain had by far the most powerful navy in the world. To finally strangle the French, royal orders were issued to prevent neutral nations—notably the United States—from trading with the French. In addition, to keep up the ranks, which had been depleted by battle, disease, and desertion, of more than 140,000 sailors needed to operate the immense British navy, the English not only pressed Englishmen out of English merchant ships and from their own homes, but also American citizens from American merchant ships.

Impressment was lawful kidnapping. During wartime, the British navy was permitted to seize any able-bodied man unable to prove he practiced a "protected" trade. Fishermen, ferrymen, iron founders, and others whose work at home was thought crucial were given letters meant to exempt them from the press. In actuality, the "letters of protection" that poor apprentices carried with them were often torn up by the press-gangs, and the men taken anyway. As the need for sailors increased, the British began to search American merchantmen for "runaways," but in practice, they took anyone they wished.

In his war message to Congress on June 14, 1812, President James Madison complained first and foremost of the "thousands of American citizens, under the safeguard of public law and of their national flag, [who] have been torn from their country and everything dear to them" and forced to become British sailors.

The point of American democracy, as President Madison saw it, was to protect the liberty of America's citizens and the freedom of her trade. On June 18, war was declared.

July 17, 1812. About two P.M.

The United States frigate the USS *Constitution* stood on the starboard tack, headed north by east with every scrap of canvas spread. The breeze was light and fluky. Capt. Isaac Hull was bound for New York, where he was to rendezvous with Commodore Rodgers's squadron in order to harass British shipping.

There was only a froth of foam at the sharp cutwater as the ship made way. High above this white apron rode her figurehead. Its maker, William Rush, described it thus: "An Herculean figure standing on the firm rock of Independence resting one hand on the fasces, which was bound by the genius of America, and the other hand presenting a scroll of paper supposed to be the Constitution of America with proper appendages, the foundation of Legislation." No god of wind or war surmounted the bow of this naval vessel. It was and remains the only naval ship ever named for and symbolized by a piece of paper.

The lookout at the masthead hailed the quarterdeck. He made out four sail of ship inshore to the south-southwest. Thinking this might be Rodgers's squadron, Hull altered course to intercept them. Even were it the enemy, he must have reasoned, his orders had specifically stated, "You will not fail to notice any British ships that you may encounter." Two hours later, another sail was descried coming from the north. As day was waning, Hull decided to make for the single ship first to see who she might be.

At about ten that night, he had closed to within a few miles of the stranger and caused the private signal to be made. The signal was kept up for more than an hour, without answer. This ship, at least, stood a good chance of being the enemy. Hull stood off to the east to await the dawn.

As the sun came up, his heart sank. There in his lee were two English frigates, with a third directly astern of him. Less than a dozen miles off, he could make out another frigate, a ship of the line, a brig, and a schooner, all in hot pursuit. Seven to one. Not good odds for a fight, and just as Hull made his plans to escape, the wind died completely, leaving him without even steerageway. The *Constitution* was not in command. Her head slewed around so as to face the pursuers, who still enjoyed a light breeze and were consequently coming up fast.

The *Constitution* was the pride of the tiny American navy. Commissioned in 1794 to help defend the growing American merchant marine from Barbary pirates in Mediterranean waters, she was not launched until three years later. She and her sister ship, the *United States,* were both forty-four-gun frigates, unusually large for that class of ship. On the other hand, at the outset of the War of 1812, the entire U.S. Navy totaled seventeen ships.

The young American nation had the second-largest merchant marine in the world, but the smallest navy of any major power. The British navy, on the other hand, constituted about nine hundred ships, half the total naval strength of the globe. Furthermore, frigates were but the light cruisers of the British navy. Its greatest strength lay in its ships of the line, huge two- and three-decked ships capable of carrying from seventy-four to more than one hundred heavy cannons.

Since naval warfare consisted of two opposing ships ranging up close alongside each other and battering each other's hulls, rigging, and personnel with cannonballs, chainshot, and grapeshot fired at point-blank range, the larger, heavier, and more heavily armed ship tended to prevail. Almost all the cannons were located on the sides of the ship—hence the term *broadside* for the full discharge of all cannons on one side of the ship— while the bow and the stern were weakly protected, if at all. When one ship was faster and more maneuverable than another, or when one combatant had its enemy outnumbered, a ship might maneuver into position to "rake" the opponent's bow or stern. A broadside fired into a warship from either of these positions could do terrible damage, both to the ship's structure and to her crew. During most of the quarter millennium in which sailing warships were the principal instrument of state power abroad, the gundecks, where the bulk of the crew gathered to fire the cannons, were painted red, so as not to show blood.

Isaac Hull must have had blood, wood, and wind on his mind as his new charge drifted out of control and pointed his eyes straight at the oncoming enemy. Once the first enemy ship came up with him, all it would have to do would be to sufficiently wound the *Constitution*'s rigging so that the ship's speed and maneuverability would be impaired. Then, all the British ships could range up alongside and compel the American frigate to surrender or be destroyed.

He had to run, but to run, he needed a breeze. The *Constitution* was a fine ship, made principally of live oak *(Quercus virginiana)* from the Georgia sea islands and white oak *(Quercus alba)* from New England. She was stiffened with the designers' diagonal braces, and Hull had recently had her bottom cleaned of barnacles. If only the wind would come up, he might yet get away. Otherwise, he would have to either send his crew to slaughter or surrender one of the two largest ships of the U.S. Navy.

Surrender was by far the more likely alternative. Indeed, in the whole era of sailing navies, the object was less to sink the opponent than to compel him to strike his colors. The vanquished ship might then go to become part of the victor's own navy. Indeed, among the flotilla pursuing the *Constitution*, one frigate, the *Guerriere*, had been captured from the French back in 1805. (French warships were in vogue among British captains, since they tended to be faster and more maneuverable than English-built ships.) The brig *Nautilus* had been captured less than a week earlier from the Americans. The *Constitution* stood fair to become the latest addition to the British commander Commodore Broke's growing squadron.

Still, Isaac Hull had two great advantages: a willing crew and an unusually well-built ship. American sailors were recruited, not pressed or drafted, into the navy. While the service was hard and desertion common, still the American sailor could count on a rate of pay five times better than his British counterparts. President John Adams had fixed salaries for American ordinary seamen at ten dollars and for able-bodied seamen at seventeen dollars per month, comparable to skilled craftsmen's pay onshore and higher than merchant seamen. And like British sailors, all American seamen shared in the prize money when they captured enemy merchantmen or warships. Furthermore, American seamen signed on for a year at a time, while conscripted English sailors might spend years without hope of even shore leave.

The *Constitution* had no trouble finding a full crew of sailors—475 in total—before she left the Chesapeake Bay on July 12. True, Hull's crew was green—just a week out of port, and some had never sailed before—but he had been drilling them at the guns and in the rigging several times each day. They were about to face one of the stiffest tests in naval history.

The other advantage was no less telling: The Americans had by far the best ship on that patch of ocean. European dockyards had been building sailing warships for better than 250 years. Though fine building oak was common throughout most of Europe, the resource had been strained to the limit not only by naval and merchant building—merchant shipping used three times as much oak as naval—but by the competing uses of oak for buildings, charcoal, furniture, bridge timbers, roads. Furthermore, the best open-grown oak flourishes in what also makes the best wheat fields. Croplands made money much faster than oak forest, so many fine oaks were cut simply to make room for fields. "The proper diminution of oak," as one writer put it, showed the vigor of the British economy. Wars were being fought in Europe for control of the Baltic trade, whose main commodities were oak plank, wood for masts, pine tar, and other naval stores. European builders had to economize, both on the thickness of their planks and on the size and spacing of the ribs that formed the structure upon which the ships were erected.

No such limitations faced Joshua Humphreys and Josiah Fox when they designed the *Constitution*. Shipbuilders usually sought either great strength or great speed in their ships. The thick sides and squat dimensions that made for the former represented a choice against the slender proportions and lighter construction for the latter. Unencumbered by traditional European habits and economies—on the Continent the French always built for speed while the English built for staying power—Humphreys and Fox went for both strength and speed. The *Constitution* was three feet wider and twenty feet longer than the average British frigate; she was a foot wider and thirteen feet longer than a standard French frigate. The *Constitution*'s live oak ribs were spaced less than two inches apart, while the ribs in most European ships were a foot apart. The close-spaced rising ribs were more than framing; they acted as an additional wall. When white oak planking was added both outside and inside, the total thickness of her hull at its strongest parts was better than twenty-two inches, where the European ships seldom managed more than fourteen inches, and that with widely spaced ribs, between which the ship was vulnerable.

The result was a ship that might well be both faster and stronger than any frigate in Europe. The heavy timbers, which would tend to slow the boat, had been offset by giving her a sharper profile and, above all, by stiffening her sides. By increasing the number of ribs and by adding diagonal braces, the American designers managed to dramatically reduce hogging. They got a frigate bigger than any other, while at the same time probably faster.

The "probably" Isaac Hull intended to try, if he could only find a breath of wind. Meantime, his enemy was closing on him, though now the wind was dying for them as well. No sooner had Hull seen the position he was in than he had all the *Constitution*'s boats put over the side. The rowers towed the ship's head around, and began laboriously to drag her away from her pursuers.

As soon as the British hit calm water, they too put boats over the side and began to row furiously toward the *Constitution*. Four of the British frigates were almost within gunshot when the American First Lt. Charles Morris had an idea. He'd just heard the leadsman call the depth at twenty fathoms, not too deep to anchor. He suggested that Captain Hull warp ship.

This was a practice seldom used except when bringing a ship to anchorage or freeing it when grounded. The idea was to tow a two-and-a-half-ton anchor tethered both to the ship and to the belly of one of the rowboats out as far ahead of the ship as possible, cut the anchor loose from the rowboat, let it take ground, and then have the sailors at the capstan in the ship's bow essentially "wind" the ship forward until it reached the submerged anchor. While this was going on, another boat would take the second anchor out as far as possible and repeat the procedure. It was as though a fisherman had cast his lure out as far as possible, snagged it on the bottom, and then drawn himself, boat, line, reel, and all, up to the snagged lure.

Why not? Around eight in the morning, the ship's boats began warping the *Constitution* across the open sea. Sailors strained to pull out the heavy anchors with the additional weight of their twenty-two-inch cables. Others strained to reel the capstan in order to pull the ship forward. They pulled with a will for about an hour, but if anything, the British were still closing.

One of the enemy frigates fired a broadside, but it fell short. Half an hour later, a second frigate tried to reach the *Constitution* with a shot, but they were still just a hair too far off.

For three hours, all hands on all ships continued the backbreaking pursuit, the Americans warping forward and the British towing up behind. The rhythm of the work must have encouraged each side alternately: When the *Constitution* had just warped up to her anchor, she must have felt she was escaping. Then, in the changeover to the second anchor, the British must have thought their steady rowing made them just about to overtake their quarry. From the vantage point of heaven, it must have looked like a race between slugs and an inchworm. By two in the afternoon, men must have been dropping from fatigue.

Then Commodore Broke had a brainstorm. He had the ship's boats from seven different ships at his command. Why not use the boats from all of them to tow the leading frigate—the *Shannon*—up to the fleeing *Constitution*? No sooner said than attempted. It is important to imagine the heart of the Americans, who saw this happening, and who just kept on warping ahead.

It was a formative moment for what would become one unvarying feature of the American character. Who could understand an outgunned, outmanned, and outdistanced frigate that would just keep on going? The sun westered and sank over the Atlantic coast, and the Americans were still warping. The evening star shone in the sky, and the Americans were still warping. British frigates were almost up to them again and again, and the Americans just kept on warping. Dinner came and went, the dogwatch came and went, and the Americans just kept on warping. Row. Drop. Reel. Row, drop, reel.

It was eleven at night on July 18. The Americans had been warping for fifteen hours without a pause. The *Constitution* was barely, just barely, holding her own against the seventeen boats trying to bring the *Shannon* up to her. Just one solid broadside from the British frigate, and the *Constitution* would be as good as caught.

About a quarter past the hour, Isaac Hull felt something on his cheek and tried to brush it away. Then, he knew it for what it was: the puff of a breeze. Quickly, he sent topmen aloft to prepare to set sail. God willing, the breeze would hold. God willing, they would get to race the whole damn British navy.

At twenty past, he gave the order to set tops and courses. It was a crucial moment. If the breeze had died at that instant, the dropping sails would have become an impediment, resisting the capstan turners and possibly slowing the *Constitution* just enough to lose her. But the breeze held. She gathered way. White foam appeared beneath the figurehead.

Quickly, Hull signaled the boats, who were about to be overtaken. They swung alongside, their crews vaulted aboard, and hauled up their boats without the loss of an instant.

The British, seeing the *Constitution* under sail, immediately followed suit. They picked up their boats and came on through the gathering dark. All that night, Hull led them a chase. At dawn the next morning, one British frigate was close enough to fire on her, but the British captain held his fire, probably afraid that in the light air the recoil of his cannons would becalm his ship.

About nine in the morning a strange sail appeared. The British, thinking it likely an American merchantman, hoisted American colors to try to decoy her into a trap. But Hull hoisted British colors, and the American hauled his wind and made off to the southwest.

All that day the *Constitution* fled, with a rising wind in her sails. A squall came up about sundown, but Hull had seen it coming, furled and reefed to meet it, then unfurled and squared away the moment it had passed. It was brilliant sailing, perfectly executed by a willing if exhausted crew.

After the squall, perhaps Hull thought for the first time, "We're going to do it. Damn it all, we're going to do it!"

The breeze kept freshening through the nighttime hours. At daylight on July 20, two full days after the chase had begun, only three sail of the enemy could be seen from the masthead, and those three were at least six miles off. Hull wet his sails to improve their performance and kept cracking on.

At eight-fifteen in the morning, the British gave over the chase and hauled their wind to the south-southwest, apparently for New York to set up a blockade of that port.

A lone American frigate had outsailed an entire British squadron. As the *Constitution*'s sails receded before him, Commodore Broke must have wondered whether this tiny American navy might be the beginnings of a formidable opponent. Scarcely one month later, he would have his answer.

Isaac Hull was daring, but he was not a fool. Immediately after his escape, he made for Boston to reprovision and to replenish the water that he had thrown over the side to lighten the boat during the chase. He waited there impatiently for instructions from Washington or from Commodore Rodgers, but none were forthcoming. When the wind shifted into the west on the August 2, allowing him to sail out of Boston harbor, he reluctantly decided to sail without orders, first writing a letter to the secretary of war, explaining his reluctance, his reasons, and his intentions. Boston was a hard harbor to get out of, the winds being often contrary, he wrote, and if he remained longer, he might find himself blockaded in by a superior British force. Nonetheless, he would not sail south to give Broke's squadron another shot at him. Instead, he'd sail northeast to the principal shipping lanes of British Canada, hoping to wreak havoc there on British merchantmen and to recapture any American ships that might have been taken as prizes and sent, lightly manned, to British harbors.

The three major British ports in Canada were Halifax, Nova Scotia; St. Johns, Newfoundland; and Quebec City, Quebec. If you imagine the Canadian coast as a mouth, St. John was on the upper lip, Halifax on the lower, and Quebec City in the throat. The Gulf of St. Lawrence was the bulk of the mouth, and the St. Lawrence Seaway the throat. By placing himself just offshore of the lips, Hull could prey on the ships bound in or out of any of the three ports. He counted on his belief that his British counterparts either were doing the same thing in American waters or were actively blockading American ports.

On August 10, he had his first success, running down a small British brig bound to Halifax from St. John. Brigs were the principal type of merchant vessel in use throughout the world. They had two square-rigged masts, sometimes with a schooner-rigged sail trailing on the main mast. They were the original tubs, meant to carry as much cargo as possible, not too quickly but very safely, with a small crew. This one must have been dwarfed by the 147-foot-long, three-masted *Constitution*. She carried no cargo of note, however, nor was the ship worth much in herself. Hull took the crew off and burned her.

A day later, he came on a bigger prize, the British brig *Adeona*, bound from Nova Scotia to England with a full load of timber. Again, he took off the crew and burned the ship. Four days after, he sighted five sail of ships. Closing on them, he found a British warship (but a mere sloop, a single-masted craft) and four of her prizes. One was already aflame when the *Constitution* approached. Two were American brigs, which he liberated. The sloop of war got away.

From the crews of these ships, he got information that Broke's squadron was on the Grand Banks off Newfoundland. If the hounds had come north, reasoned Hull, the fox would head south. He set his course for the waters off Bermuda, where he expected to harass British ships entering and leaving the Caribbean.

On the night of August 18, he chased another brig, which turned out to be the American privateer *Decatur*. So concerned was her captain to escape the big ship that loomed up on him out of the dark that he had pushed overboard twelve of his fourteen cannons. To no avail. The *Constitution* brought him to after less than an hour's chase. The captain told Hull that the day before he'd seen a large warship to his south, and that she could not be far away. In fact, it seems that the *Decatur*'s captain had thought the *Constitution* to be that ship come in chase of him.

The *Constitution* continued her course southward, hoping to fall afoul of this supposed warship. At two in the afternoon on August 19, running before a steady north wind, Hull got his wish. He was in midocean, about due south of Cape Race, Newfoundland, and due east of Boston, 318 miles from the nearest land. The lookout at the masthead cried a sail to the south-southeast, too far away to say what sort of ship.

Hull squared away and gave chase. An hour later, they were closing fast. Their quarry was a large ship under easy sail on the starboard tack, making no effort to get away. By three-thirty, it was plain that this was a large frigate, and not an American. In fact, it was the *Guerriere,* lately part of Broke's squadron, but now cruising alone. Her captain, Richard Dacres, was a distinguished second-generation British naval officer. There were impressed American seamen aboard the *Guerriere,* and when they saw what was afoot, they earnestly requested of Captain Dacres that they not be compelled to fight their own countrymen. Part gallant and part practical, Dacres sent them below as noncombatants. The last thing he would need in a close action would be crew members who might suddenly fight for the other side.

However fast the *Constitution* might be, it appeared Dacres had no doubt he could beat her. He backed his main topsail to check his way, clearly inviting the *Constitution* to come closer. Hull responded by sending down his royal yards, taking in staysails and flying jib, and taking a second reef in the topsails. This kept him on course to intercept the stranger, but at a slower pace. He hauled up main and foresails and beat to quarters. The crew gave three cheers and manned their guns.

As they came up, they could see painted on the *Guerrière*'s topsails the words "Not the *Little Belt*." The *Little Belt* had been a British brig of war defeated by the *President*, the third American forty-four-gun frigate, a year before, in a fight of unequals. The implication of the words was clear: In a fair fight between equals, His Majesty's ships would always prevail.

Dacres was cocky. He turned to try to get to windward of the *Constitution* so he could dictate the moment of battle. Failing that, he hoped to get a broadside that would rake her from the bow or stern, doing great damage immediately and reducing the likelihood of carnage on his own ship.

As Hull came up on his weather quarter, Dacres let fly a premature broadside that fell well short. He turned to give the *Constitution* his larboard broadside. This time, he succeeded only in shattering the *Constitution*'s knighthead—a five-inch-thick piece of live oak on the quarterdeck, largely decorative— with one ball and cutting in two a fish hoop on her foremast with another. The Americans responded only individually.

The fight between USS *Constitution* and HMS *Guerriere*
(Courtesy of the Naval Historical Center)

Unable to get the weather gage, Dacres turned to give Hull the opportunity to come up alongside him. Hull might have wondered at his adversary's nonchalance, but he did not let the chance slip. He struck his jib, backed his main topsail, and slowed his way. A few hundred yards off, he set the main topgallant to propel him in close and came alongside the *Guerriere* at a distance of less than a pistol shot.

It is said that Hull's crew asked three times to fire a broadside and were refused. At 6:05, Hull consented. The evening erupted in smoke, fire, and flying metal.

A ball went straight through the *Guerriere*'s hull, right between wind and water, where a hole would let water in. Another went through not five yards away.

The *Guerriere* generally fired higher, perhaps hoping to take this prize with as little damage to the valuable ship as possible. The first few broadsides shot away the *Constitution*'s main braces and put holes in the main, fore, and fore topgallant sails.

One ball struck the *Constitution*'s twenty-two-inch-thick white- and live-oak sides, briefly hung there, and then fell into the sea. One of a nearby gun crew is said to have seen this and shouted, "Huzzah, boys! Her sides are made of iron!" giving the ship the nickname she has borne ever since, "Old Ironsides."

Ten minutes later, the *Guerriere*'s mizzenmast, shot through, toppled into the sea, its weight dragging the boat to port. The *Constitution* continued pouring in steady fire at a rate of one broadside every two minutes, which had so far put more than thirty shots into her opponent's hull between wind and water.

Hull tried to turn the *Constitution* for a raking shot, but with his damaged rigging he could not manage it. Dacres in turn tried to take advantage of his dragging mizzenmast to let him turn behind Hull and rake the *Constitution* from the stern. But the mizzen so accelerated his turn that his bowsprit ran straight into Hull's cabin on the quarterdeck, becoming hopelessly entangled. It was only the whole weight of this ramming frigate that actually succeeded in doing any damage to the *Constitution*'s hull, and this damage was largely cosmetic.

The *Constitution*'s gunners kept up their fire, now shattering the *Guerriere*'s larboard bow and tearing a six-foot-long hole in her planking. Both sides had tried unsuccessfully to board the other when the *Constitution*'s greater momentum tore the ships apart. The shock to the *Guerriere* caused her wounded foremast to fall, and the foremast pulled the mainmast after it.

The British frigate was now neither a brig (two-masted) nor a sloop (single-masted), but entirely dismasted. Helpless, she rolled on the waves, taking on more water through the battered hull with every roll.

Half an hour later, having hauled off to get his rigging back into some sort of order, Hull approached the rolling hulk of the *Guerriere*. It was now full dark, and the Americans could not tell if the captain had struck his colors, so Hull sent a boat aboard her to find out. The boat returned with Captain Dacres, who had indeed surrendered.

Though Dacres was later, with some justification, to blame everything from rotten masts to bad luck for his defeat, it was clear that his early cockiness had been unjustified, indeed perhaps his downfall. He had not bothered to calculate either the quality of the American ship or the spirit of her crew.

The Americans tried to get the valuable *Guerriere* under tow to take her in as a prize, but their gunnery had been too devastating. The water in the hold was five feet deep and gaining, in spite of constant exertion at the pumps. (Sailing ships were never entirely watertight, since the oak planks worked in a seaway. A ship was generally pumped out three times each week as a matter of course.) The next morning, Hull ordered everyone off the dying vessel and set her afire. When the flames reached the powder magazine, she blew to bits.

In disbelief, the London *Times* reported, "Never before in the history of the world did an English frigate strike to an American."

But before the year was out, two more British frigates would be soundly defeated by American frigates, including the *Java*, again by the *Constitution,* this time under Capt. William Bainbridge.

The British admiralty—among the most powerful government institutions in the world—was panicked. Fresh from the overwhelming British victory at Trafalgar, they had run straight into unexpected defeat at the hands of one of the world's smallest naval powers, their own erstwhile colonies. The admiralty issued orders that no British frigate was to engage an American forty-four in single combat. Their confidence had evaporated.

Thereafter, the British admiralty showed a constant interest in American shipbuilding. In response to the forty-four, they scrapped a whole class of small British ships of the line and adopted ships with scantlings closely akin to the forty-fours. British spies kept watch on powerful ships of the line under construction in the Philadelphia shipyard. When in 1816 the

Franklin—a new American ship of the line mounting eighty-two guns that could throw 2,624 pounds of metal, as opposed to an average British weight of a little under two thousand pounds—visited England, furious correspondence went back and forth between the admiralty and the comptroller of the navy. Had they ships, they wondered, to counter the American threat?

End of the Age

Oak saw the nations of the West to the dawn of a new age. And there it left them. Oak was indeed not adequate to the ambitions of the modern states and their citizens. Large oak ships with cavernous holds were fine and long-lived for as long as they were in use, but when they were left unused and unventilated in peacetime, fungi infected the hulls, particularly in the crucial timbers "between wind and water." Sometimes wet and sometimes dry, these timbers were in just the right spot to promote the rapid increase of wood-decaying organisms. Samuel Pepys complained of inspecting such ships "laid up in ordinary" and picking up handfuls of toadstools as big as his fist. When an inspection by an admiral was planned, it was often necessary to send men ahead to shovel the fungi out of the hold.

During the Seven Years War alone, sixty-six British naval ships set sail, never to return. They were not lost in battle. They were simply lost. In all likelihood, they sailed off their bottoms—the rotten timbers between wind and water giving way in heavy weather—and sank like stones.

The *Royal George* lost its bottom at anchor as it was being heaved over for minor repairs. A one-hundred-gun ship of the line, she was among the most celebrated British ships. She had been Hawke's flagship at Quiberon Bay in 1759. In 1782, she was to be the ship of the popular and capable Admiral Richard Kempenfelt. The admiral himself was aboard with several hundred crewmen and a number of distinguished visitors when the heeling over began. At that moment, the hull gave way and the ship capsized, drowning more than eight hundred people, including the admiral.

Songs were written about the loss. It was a national tragedy, when the bottom fell out of the *Royal George*. The phrase that survives today describes not ships but stock markets: "The bottom fell out of the market." For all their pretension to stability, both the oak ship and the state economy were subject to sudden disaster.

The age of oak ended on March 8, 1862, at Hampton Roads, Virginia.

Not long before, the forces of the newly minted Confederate States of America had rushed to take over the shipyard at nearby Gosport, Virginia, from fleeing Union troops. They did not arrive, however, before the Yankees had set fire to the most valuable ships in port, including the famous *Merrimack*, pride of the newest class of large, steam-driven warships. This ship had so impressed the British admiralty on her visit to England that the British immediately commissioned a class of similar ships.

The *Merrimack* settled in shallow water, with her engines and lower hull intact. Within four months, the Confederates had raised the hulk, meaning to use it as a platform and power plant for a steam-driven warship whose sides would be made of iron. The British and the French had already developed such ships, but the Union navy had not even one. The CSS *Virginia*, as she was dubbed, would beat anything the federal forces could throw against her and, above all, would break the blockade that prevented the Confederacy from receiving the supplies that it needed in order to fight.

Hearing this, the U.S. Navy immediately set about making their own ironclad. Time was short, since the rebels had a head start of some months. The navy settled on a novel design created by an ambitious engineer, John Ericcson. A Swede by birth and the son of a mining engineer, in his peripatetic career he had already designed everything from improved sawmills, to steam engines, to locomotives, to screw propellers, to depth finders. His clients included Swedes, Englishmen, Frenchmen, and Americans. In fact, he'd already tried to interest Napoleon III in his "impregnable floating battery," without success. Now the U.S. Navy bought it, a ship with more than 60 percent of its space below the waterline and with a revolving turret mounting two guns amidships. It looked like an iron apron with a large wart, but it could be built quickly and the navy hoped it would prove as "impregnable" as advertised.

On the morning of March 8, Confederate Commodore Franklin Buchanan hastily put his crew aboard the finished CSS *Virginia*. The ship got up power and steamed down the Elizabeth River toward the blockading Union ships at Hampton Roads. She had iron plates up to four and a half inches thick covering her from stem to stern. She looked, remembered her chief engineer, "like the roof of a barn afloat." Under full steam, she was making nine knots, not enough to catch a fleeing sailing ship with a fresh breeze but more than enough to deal with blockading ships trying to maintain their stations.

The chief ships of the Union squadron were both sailing vessels made of oak, the frigate *Congress* and the ship of the line *Cumberland*. Buchanan steered between them. Fearing that the new rifled guns of the *Cumberland* might penetrate his iron, he steamed right for her and rammed her near her starboard bow. The wrenching force tore the iron prow off the *Virginia*, doing no structural damage, but the broadsides from the *Cumberland* had done no damage at all. There was a gaping hole in the *Cumberland* and she began to sink, still firing ineffectually.

The *Congress* too had been firing broadside after broadside into the *Virginia*. Now the ironclad returned fire. Within half an hour, the frigate was a wreck, and she hoisted a white flag. When Union sharpshooters onshore fired on the Confederate vessel, Buchanan resumed the fight, firing heated shot into the *Congress* until she caught fire. All night she burned, until just after midnight, when the powder magazine caught fire and the frigate exploded.

The next day, the reason for Buchanan's haste was clear. Into the estuary came the new Union ironclad, looking for the still virtually unscathed *Virginia*. The two fought an indecisive battle for about three hours. The *Virginia*'s bow sprung a leak after she tried to ram her opponent, and the Union ironclad's revolving turret was slightly opened, as though by a can opener. Both

End of the age (Courtesy of the Naval Historical Center)

retired, but the *Virginia* now knew she could not strike the wooden ships with impunity so long as the *Monitor* was near. Neither side had won, but both had a new arms race to begin.

Six months later, the Union ironclad went down on New Year's Eve in a storm off the North Carolina banks. Though "impregnable" enough, she was not very seaworthy. She left with us her one battle and her name, conceived by her engineer creator: *Monitor*, which in Greek means "the warning."

OAK ITSELF

I WENT TO visit a local sawmill. It was late winter. The loggers had been busy, cutting and skidding butts of oak and maple, hemlock, ash, and pine off the steep slopes, selling them at the mill. The yard was full of thick butt logs, separated by species, roughly piled.

It was icy underfoot, but the ice was starting to melt in places, making for dangerous walking on the rutted ground. The trunks varied in diameter between eight and about thirty-six inches, and the sections were between six and twelve feet long. The oak pile was the largest. It was mainly red oak *(Quercus rubra),* with some white and chestnut oak *(Q. alba* and *Q. prinus).*

A big yellow payloader picked up a bucketful of oak logs, carried them to the mill, and dropped them onto a bed of rollers that stepped them down an incline until they reach a sluice.

A log shot forward and rotary grinders—enormous steel

cylinders full of hobnails—ground off the bark. The naked log was then shunted onto another set of rollers, in line for the saw.

The log fell onto the saw carriage. Steel saw dogs gripped it. The operator set the dimensions, and the log began to rocket back and forth across the upright circular saw blade, turning with each pass.

First the sapwood came off. It shot down a conveyor belt where a man unloaded it into a pulp machine. Then the saw started to take off boards and fire them ahead. The same man shunted them onto a field of rollers where another man roughly graded the boards: #1 common, #2 common, #3 common, #3B common, or one-and-better, select-and-better. When the machine had reduced the log nearly to the pith, it cut and spit out a four-inch-square cant.

The log was gone.

The whole process took little more than a minute. The next log was already on the saw.

Number 2 and 3 common oak boards are 50 percent or less clear wood. The rest is knotty or has other flaws. It is sent to plants where it is made into flooring. Number 3B—even knot-tier—goes to separate plants, where it is made into flooring for box trucks and trailers. The cants go to plants that manufacture packing palettes.

Number 1 common oak boards are two-thirds clear. Select-and-better is 85 to 100 percent clear. These are for the cabinetmaker. They go to manufacturers of kitchen cabinets; the very best boards to custom cabinetmakers.

It does not matter to the high-speed saw whether the wood it cuts is oak or ash, maple or poplar, hemlock or pine. Given the chance, it would try to convert an icebox into boards.

Industry did to craft what Babylonian agriculture did to gathering. It caused a tremendous increase in output, but at high human and environmental cost. Industry often starts as a liberator but ends as a slaver because it cannot control itself. Its solution is to make more and to make it faster. When sooner or later it falls into the hands of the inept or the greedy, they run it full tilt, heedless of the consequences until the soil is salted, the Dust Bowl blows away, the forests decline under acid rain, the market crashes, Bhopal suffers under a poison cloud, Chernobyl melts down.

But who would want to do without the brilliant high-speed saw? It does away with so much backbreaking labor. The waste of energy a man encountered after half a day of cutting and splitting when a log turned out to have a ring shake or a black heart is no more. Though one might still want to make cleft oak boards, one might not want to do it for a living. It is tedious and hard.

Is there no way to have the fast, beautiful tool without getting the headlong rush, the anxiety, the ennui, the cruel trampling of surplus people?

Human beings show restraint when they value, worship, and respect what they encounter. Value comes from understanding, and understanding from intimacy. Humans in the age of oak had to confront the resistance of their materials every day. Memory, reason, and skill wove a world of oak. The people understood, valued, and indeed worshiped the tree that was their intimate companion and source of so much of their livelihood.

The bread, the tracks, the henges, the churches, the houses, the roofs, the ink, the casks, the leather, the ships of oak are mostly gone now. Oak is for truck floors and middle-market cabinets. There is a small but growing market for fine oak furniture in the arts and crafts style, which shows the beautiful wavy grain of quarter-sawn oak. The buyers are mainly cultivated urbanites, nostalgic for honest craft and simpler times.

But oak is not forgotten. Natural scientists, ecologists, and engineers remember it. They do not break down the oak tree into its genetic components, then recombine them and patent the results. Instead, they closely examine the tree in nature, to learn from its own life history and its own design.

In the 1970s it occurred to scientists that the annual rings in oak trees could tell time. Formed each year during the growing season, these rings vary in thickness according to environmental conditions. A cold year or a drought year, for example, will produce a much narrower annual ring than a warm year or one with abundant rainfall.

Archaeologists realized that the annual rings formed by different oaks in the same area would have similar patterns of wide and narrow rings, because they had experienced similar conditions. If they could learn to read the codes in the rings, they could in theory start from oaks today and—link by link, matching pattern to pattern—count back toward the beginning of humanity. The assembled patterns of the oak chronology would then allow the archaeologists to date precisely any oak artifact found at a site.

Chronologies are actively being created in Ireland, England, and Germany. Among them, they now allow dating that goes back almost ten thousand years, to the very beginning of the Holocene. And unlike radiocarbon dates—which at best have an accuracy of plus or minus ten years—these dates are exact. A good dendrochronologist, as the scientists in this new discipline are called, can even say in what season of the year the wood was cut.

What use is such exactness? For one thing, it will allow students of the past to know if neighboring archaeological sites were really neighbors or if one succeeded another. It has also begun to produce unanticipated help in interpreting the environmental history of the Holocene. Some narrow ring patterns found in many of the existing oak chronologies correlate with probable major volcanic eruptions and records of darkness, famine, and other portents, from as far away and as deep in time as the Shang dynasty in China, roughly 1750 B.C.

Recently, too, close observation of the natural laws of tree growth has begun to provide models for design. No tree is able to hold out heavier and longer branches for as long and as safely as an oak. Klaus Mattheck studied oaks and the other trees in German forests to learn how they could do it. From his observations, he postulated the Axiom of Uniform Stress, which states that trees actively grow to equalize the stresses over their entire surface. Wherever a dangerous stress point occurs, the tree grows new wood to control and reduce it.

Mattheck cleverly invented a way to simulate this adaptive growth on any object and modeled it. The method—called Computer Aided Optimization (CAO)—reveals the high-stress points in any designed object and treats them as though they were the parts of a living tree. Like a tree, the program "grows" new tissue for the component at the points of high stress until the dangerous weakness is eliminated. The product is then manufactured using the growth data. The sort of surgical screw used to mend complex bone fractures, for example, is now routinely made using this method. It lasts twenty times longer than did earlier products.

The study of oak itself, then, is a school of history, design, and society. There is no structure more supple and sustainable than nature's, and oaks are among the most widely adapted and successful of all plants. Close attention to oak—the way it is built and the way it works—may lead to more sustainable designs for everything from clothes to transportation systems.

Just as students of early humans discovered that people did not in fact hunt big game to extinction in order to establish mankind, so those who study oak are finding that the tree's success is owed not to the ruthless suppression of competitors but to its flexibility.

Diversity

The history of our planet is the story of radical changes in landscape and climate. Continents split, moved, and collided. Temperatures, seasons, rainfalls, and ice sheets rose, fell, appeared, disappeared, shifted. The botanist Edgar Anderson referred to such changes as "hybridization of the habitat."

It was the great virtue of oaks that they responded not by specializing and narrowing their range, but by adapting, expanding, and radiating into more and wider-flung landscapes. There have seldom been creatures so tenacious as oaks, but their staying power is founded on their own ability to change.

If you suggest to any two oak taxonomists today how many species there are, you will start a fight, because they will never agree. Some people think there are 450, others 250, and most think there are any number in between. Some say that species are really hybrids or hybrids really species. But even we casual observers can tell them that there are lots and lots.

All oaks have acorns, but sometimes the acorn is about the size of a .30-06 rifle shell, and when it falls from the tree on your head, it feels like one. Others are fat and round like cherry bombs. (A friend who lives in a red oak wood outfits his kids with batters' helmets when they go out to play in autumn.) Some acorns are so small you could mistake them for martini olives; others look like dark brown Christmas lights.

The caps that attach acorns to their stems are equally various. Some are set on the seed like a beret two sizes too small; others are like modest grass skirts that cover practically the entire fat delicious pod; others are frilly and revealing. Some are hairy, some are scaly, some are knobbly, some are woolly. Some resemble the helms of kings, others peasants' caps.

And what of the trees themselves? One's crown is 120 feet wide and almost as high. Another is a single pillar rising ninety feet, with not one branch until sixty feet. One never grows higher than your knee but spreads like crabgrass; another

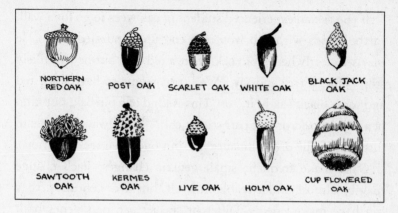

A few acorns (Nora H. Logan)

throws its top branches skyward and its bottom branches earthward like a dancer caught in two poses at the same time. One lives in swamps where crawdads dig their burrows among the roots; another lives in uplands where there are prolonged dry spells. One has leaves as small as your thumbnail, another as large as a napkin. Some of the leaves are spiny and hard like the sloughed exoskeleton of a beetle. (If you sit on one, you get right up again.) Others are as fine and delicate as bible paper. Some hold their leaves for years, others hold them from spring to spring, and some lose them every fall.

But for all that there are many species of oak, each one differs comparatively little from its near relatives. Among the genus *Quercus,* only a few genes can make the difference between one species and another. Out of seven hundred genetic markers, for instance, only six distinguish *Quercus grisea* from its neighbor, *Quercus gambelli.* Oaks are very adaptable, because small genetic changes can make noticeable differences.

If you ask an experienced student of oak trees to go for a walk in the woods with you, you may end up as puzzled as he. He may see clearly here a pin oak, there a red oak, but he will likely stop before a tree and say, "Well, this seems to have some red and some black oak in it," or "This is kind of a pin oak, but kind of a willow oak, or is it part scarlet oak?" It is always possible to distinguish the predominant type, but one often sees variations.

Oaks make frequent small genetic changes. Each change may be almost unnoticeable in itself, but over centuries, they can have major effects. Different species can pass genes back and forth, and through generations of crosses and backcrosses, new characteristics—both in the habit of the tree and in the habitat it prefers—come to be.

Often, this ability seems mere genetic doodling. A slightly scarlety pin oak seems to have no advantage in the landscape over a slightly willow-oaky pin oak. But the process is powerful over the long term. Many people think that when two species share genetic information the result will be a plant whose characteristics are halfway between the two. This is not necessarily so. A recombination of genes can also produce wild distinction, children with little resemblance to either parent. Most will not even be viable. They may want a nonexistent climate, or they may be sterile. But a few would be viable, if only they had the opportunity.

The few viable alternatives usually live out their lives and cross back into the next generation by sharing their pollen or their seeds with their parents' species. This makes for little external change—just enough to annoy or delight oak walkers in the woods—but underneath, at the genetic level, a mosaic is building. Molecular-level studies of oak populations show that even apparently identical trees often have chloroplasts containing the DNA of different species. When major environmental changes happen around oak trees, this genetic richness allows them to adapt quickly. These genetic self-experiments perpetuate a fund of possibilities, waiting a place to be used.

Tenacity

Around sixty-five million years ago, at the beginning of the Paleocene epoch, a nut fell to earth somewhere in what is now Thailand. A small hairy creature that looked like a cross between a squirrel and a rat worried the nut between its paws. It carried the nut away and buried it in a clearing. A long-lived member of the Fagaceae—the beech family, which includes all the oaks—will make at least three million acorns in its lifetime. This squirrel-rat had buried about two hundred that season. And it never again found that particular one.

Water penetrated the seed coat and activated the miniature plant waiting within. Using the food stored in the seed, the radicle—the proto-root—penetrated downward through a fissure in the bottom of the nut. The radicle senses and follows gravity, while sensing and fleeing sunlight. Turn the nut upside down in the ground after the radicle has sprouted, and it will reverse direction to keep growing downward. This first root spiraled more than a foot into the soil during the plant's first year. Already it was doing more in the way of roots than its ancestors did.

Otherwise, for about twenty years, the tree pretty much resembled its forebears. When it finally became a sexual adult, however, there was a startling change. When the male flowers came out, instead of standing up on the stems, ready to receive insect visitors, they hung, indeed they drooped, and waved in the wind. This tree, unlike its progenitors, had reverted to the more "primitive" wind pollination favored by the gymnosperms. And the nut, instead of being fully defended by a spiny husk, was open and smooth, with just a little cap to join it to the stem.

It was the first oak.

And in one regard, it was really disappointing.

Such a magnificent tree ought to have had a beautiful flower, at least the equal of the cherry or the magnolia or the tulip tree. But no. Fine flowers are for plants that need to attract, massage, and reward insects. The male oak flower looked like a few strands of kinky, limp thread. Today, most people don't even know it is there. (When the flowers fall, a few of my clients complain about the "droppings.") Even the pollen is lightweight and inconspicuous. And the female flower, hidden in the angle between twig and stem, looks like a pimple. It is hard to believe that something so large and long lasting could come from such a nondescript beginning.

Nevertheless, the male flowers made good use of the wind. They ripened, filled with yellow pollen grains, and then released a golden rain that floated for miles and miles through the countryside on the wind's convective whorls. With luck, they would strike the stigmas of the tiny female flowers hidden in the leaf axils and new acorns would begin.

From this single beginning on the edge of a proto-continent that included present-day Thailand, the oaks began their travels through changing landscapes and through generally worsening weather.

Land masses spun apart and undersea ocean ridges—the sutures of tectonic plates—spread. Pressure built up on coastlines. Mountains rose. In the shadow of these mountains, it was drier and the winters were a little colder. It was not a good climate for dependable insect pollination. If it were too cold or too dry, there might be a poor hatch, then a poor crop of bugs, then poor fertilization of the flowers. Wind operated under any circumstances and assured that at least some fresh acorns would be made. The oaks spread across Asia within ten million years.

But to the west—where Europe and North America had once been connected to Asia—the forests were blocked. The ancient Turgai Sea stood in their way.

In the middle of the Eocene, a bridge opened across the Turgai when moving continents brought two peninsulas together. The oaks streamed across and into Europe. There, although the Atlantic was becoming an ocean, a mosaic of bridges still connected what would become Europe to what would become Scandinavia, Greenland, and North America. These bridges themselves opened and closed as the lands jostled, but the oaks moved into North America.

As late as forty-five million years ago, it was quite warm in the middle of the world. The zone of the tropical rain forests stretched as far north as the present northern boundaries of California, at forty degrees north latitude, and even into coastal Alaska. Broadleaf evergreens, including the oaks, were dominant.

Today, there are more than 130 species of oak in mountain-
ous northern Mexico. It is a world full of oak. If you stand in
the middle of them, you will get a good idea of what the world
was like in the Eocene epoch, for all these oaks or their near relatives once lived far abroad across every continent. In fact, the
fossil leaves of oaks found as late as the Eocene in Italy are precisely the same as the leaves of oaks that live today in southern
California and northern Mexico.

As the Eocene ended, about thirty-seven million years ago,
pieces of the world started banging into each other again. The
Pacific Rim pushed against North America, the number of vol-
canoes increased, and the Rocky Mountains began to rise. The
sea that had divided one half of North America from the other
drained away and dried. Meanwhile, halfway around the globe,
the Indian subcontinent at last slammed into mainland Asia
and pushed up the Himalaya. In southern Europe, the Alps
began to rise.

Suddenly, about thirty million years ago, the most dramatic
cooling that the earth had known in the past one hundred million years occurred. It was a disaster for much of what had
come before. It was also the foundation of the modern world.
When it was done, the mean temperature of the globe had
declined by almost thirteen degrees Celsius.

A permanent ice cap began to cover the earth's high lati-
tudes. The oceans for the first time acquired cold-water depths.
Cold and dry seasons became longer, more pronounced, and
more unpredictable. The cold flowed quickly down the troughs
made by north-south mountain ranges, such as the Appalachi-
ans and the Rockies. It was slowed where it met east-west

ranges, such as those of Asia and southern Europe. The Turgai Sea at last disappeared for good, and Europe was firmly connected to Asia.

Insect pollination, which had seemed such a sure thing, now was less secure. A lot of pollinators died out altogether, while the survivors depended on the mercy of the weather and tended to remain where the climate stayed most like it had been in the old days. Some plants—including the oldest oaks—took refuge in the places where the climate changed little. But new oaks appeared.

In North America, where the cold penetrated farther south, the oaks evolved a new strategy. If the climate were too unpredictable to make sure that an acorn would mature in one season, why not give it two seasons?

The red oak tribe came into being. Today, all true oaks belong either to the white oak or the red oak group. The white is older. Its acorns are fertilized and mature in one season. The leaves of its member species are smooth at the tips of the lobes. Red oak species have leaves with spiny tips and take two seasons to mature an acorn. In the first year, the wind-borne pollen grains reach the stigmas of the tiny female flowers in early spring. Each grain bores down the stigma toward one of the six ovules hidden deep inside. But as the first autumn sets in, the pollen stops partway down the tube. Next spring, soon after it warms, the winning grain reaches its ovule immediately and the acorn begins to ripen. It would be hard to imagine a better way to beat a climate that was progressively less predictable.

After the great cooling, the weather briefly warmed, but soon it cooled again. In the north, it got too cold for a tree to be a broadleaf evergreen anymore. The winter drying of leaves and the burning caused by cycles of freeze and thaw made evergreen leaves (not needles) inefficient. So the oaks changed again. For the first time, some became deciduous, losing their leaves and hardening off their tissues in preparation for the cold season of each year.

This was a risky change. It meant an energy economy that had to be strictly regulated. Now there was a period of the year during which no photosynthesis occurred, and the stored energy of the previous year would have to be sufficient to pump out a whole new crop of leaves each spring. During the time immediately following leaf-out, the tree would be severely energy depleted. An attack of pests or disease at this time, or the failure of the sun to shine, could have disastrous results.

In the last million years, the climate has had at least two dozen sharply defined pulses from warm to cold and back again. Within these, perhaps six have been deep and prolonged, lasting from forty thousand to one hundred thousand years each. At the low points, glacial ice could form at elevations more than three thousand feet lower than it does now, and masses of ice swallowed the entire landscape of the Northern Hemisphere down to about the level of Seattle, St. Louis, and New York in the United States; Amsterdam, Warsaw, and Berlin in Europe; and Kiev in West Asia. At the high points, the climate was more or less as it is now, sometimes even a bit warmer.

With each advance of ice, the oaks retreated south into places like the Mediterranean basin, the valleys south of the European Alps, Southeast Asia, and southeastern and south-

western North America. With each warming, they returned
northward. The peaks of warmth in the interglacial eras were
steep and narrow, and so were the troughs of cold in the ice
times. No interglacial era in the last million years has lasted
more than twenty thousand years.

The depth of the most recent ice age was nineteen thousand
years ago. The rolling glaciers scored rock, broke it to pieces,
and rolled the hunks beneath its weight until the rock was
ground to the texture of flour. Even beyond the reach of the
glaciers proper, repeated freeze-thaw episodes shattered stone
and left fresh shards exposed to weathering.

Twelve thousand years ago, the climate began quite suddenly to warm. Fresh, completely unleached soils liquefied
beneath the melting glaciers and flowed over the landscape.
Everywhere that there was a slope water flowed, bearing new
soils in its wake. Swirling winds lifted up the lighter of the
dusty stuff and carried it hundreds of miles, spreading it over
the landscape. This was the stage onto which humanity
emerged, among the tenacious oaks.

Cooperation

A jay appears out of nowhere and lights on the branch of
an oak tree. It preens a little, looks about. It is October,
a windy day with long tattered clouds streaming overhead, and
a great meal of thousands of acorns is ripened on the branches
and scattered on the ground beneath the tree. The jay looks
over its shoulder, then picks an acorn. It shakes the nut, turns

its head a moment as though reflecting, then drops that acorn. (A weevil had already eaten most of the seed inside. It was too light and it rattled.) The jay looks around once more, hops, seizes another. Holding the acorn securely in its long, bendable toes, the jay pushes its hooked upper bill through the shell. It clamps down with the lower jaw, then presses, twists, and tears until the shell comes off. Then it chomps and swallows the rich seed.

As if on cue, more jays appear in the area and make for this wonderful tree. As soon as he sees them, the first jay changes his mind about eating. He explores the branches, testing for sound acorns. When he finds a good one, however, he swallows it whole. This is a startling thing to see for the first time, since no one would think that such a small bird could swallow such a big thing without choking to death. In fact, he is not swallowing it all the way down. He is storing it in the expandable throat that all jays have evolved for this purpose. A jay can store up to five acorns—depending on their size and his—in this container. He stuffs in as many as he can fit, takes a final acorn in his beak, and flies away.

The other dozen jays who have found the tree begin doing just the same thing. Test, select, swallow, select, swallow, test, yecch, select, swallow, get one more, hold, fly. Each jay makes many trips to collect the fruits, harvesting both in the tree and on the ground.

While many acorns remain, the gathering is amicable and practically silent. Each bird is occupied with his or her task. When only a few sound acorns remain, however, mayhem ensues. Jays fight over the last acorns, screaming, and when the victor finally flies away in triumph, he is often pursued by more than one bird, intent on later theft.

The tree is left alone, the rich supply of thousands of acorns reduced to a few hundred, almost all of them unripe or already parasitized by weevils or fungi or both. It is quiet again, except for the sound of the wind rattling the drying brown leaves.

The oaks evolved about sixty-five million years ago. So did the jays. There are still more different species of oaks in places like Mexico's Sierra Madre than in any other. Ditto for the jays. Jays and oaks are a couple. Essentially, the oaks have domesticated the jays, or vice versa.

The list of animals and fungi that feed on one part or another of the oak would fill a small phonebook, but most just take what they need. The oak is their host, accustomed to bearing a lot of hungry guests. Indeed, every part of the oak is filled with a bitter substance, tannin, meant to control the number of guests or at least reduce their appetites. Some give as good as they get—certain fungi give water and nutrients in exchange for the food they take, and squirrels act on oaks much as jays do, spreading and planting acorns, eating maybe one for every three they bury and lose. But oaks and jays have been key to each other's survival since their beginnings and so to each other's spread around the globe.

Jays and oaks make a strange couple. We are accustomed to admire the oaks, but what about the jays? In one place, they are gray blue and look permanently dusty. In another, they are blue black with pointed crests that make them look in flight like animated arrowheads. A third group have shoulder stripes like lance corporals, and still others are a nice orangey-tan color decorated with blue sleeves. But wherever they may be and however they may appear, jays always make too much noise, an insistent, raucous, grating caw. On the edge of a wood, it is almost impossible not to stray into jay territory, and whenever you do, you will certainly hear about it. Jays screaming is the aural equivalent of mosquitoes biting. Indeed, the European jay has the genus name *Garrulus,* that is, "talks too much." But if we are going to admire the oaks, we had better learn to admire jays too, for between them they have changed the surface of the earth.

Jays do the one thing oaks can't. They move. All the time, it seems, they move. When we are passing through their territory, their flights look flitting and purposeless, though likely they are meant to distract and annoy us, to make us go away. But if you settle down and stay awhile, you may begin to see what they are doing.

The jays carry acorns away, usually to areas near where they are nesting. There, on the edge of the woods or in any case on some edge—it may be between wood and field, between tall grass and low, between garden and lawn—they bury their found acorns, not in groups of nuts, but one at a time.

The jay pushes each acorn into the soil as deep as it will go. Then it hammers with its bill until the nut is fully covered. Sweeping side to side with its bill, it covers the nut with soil and sand. It then picks up a few leaves, twigs, or pebbles and artfully scatters these over the hoard.

The average jay will bury more than forty-five hundred acorns each autumn. They are the adults staple food through the winter months, and serve both adults and fledglings in the spring. But how easy can it be to find thousands of nuts that were hidden weeks or months before?

I. Bossema studied the European jay intensively, concluding that jays buried on edges largely so as to have landmarks by which to locate their caches. Furthermore, they buried the acorns one by one—not in a single cache—so that if a scavenger found one, it would not necessarily get them all. That they could indeed find the nuts again was first demonstrated to Bossema by chance in his own garden as he reported in his 1979 study.

In the middle of October I observed a marked jay hiding food in my garden. This bird had three acorns in the aesophagus and carried a fourth one in the bill. It concealed the fruits on four different sites: the first one close to the trunk of a tree and the others along the edges of the lawn. I did not observe this jay at this place again until 72 days later. It landed in front of the hiding place close to the tree trunk and located the acorn precisely without hesitation. It flew into the tree and dehusked and ate the acorn completely. Thereafter the bird flew to the lawn, approached the edge, dug briefly twice within [a few inches] of a hiding place and found the acorn at the third attempt. After dehusking and eating it in the same tree, the bird disappeared and was not seen again in this part of the garden. In May the remaining hidden acorns produced seedlings.

Even the young seedlings are of use to some jays. When the sprouts emerge, jays may lift them, shaking them in their bills and dislodging the food-rich cotyledons that persist underground in what remains of the acorn. This procedure often kills the seedling, but as often it does not. In the spring, what's more, jays often feed their newborn on a nice pulp made from the larvae of oak leaf–eating insects.

The oaks make staple foods for the jays and have done so for epochs. Jays have changed both physically and mentally to adapt to this way of being. The V-shaped hooked upper bill is adapted to tearing husks, the expanding esophagus to storing acorns for transport. Jays have evolved a lower bill that is attached to their headbones so that the bill cannot be dislocated when hammering. And jays' breeding is timed to coincide with periods when food will be available from the oak. Mentally, jays have developed an extraordinary ability to remember landscape features to be able to find again the acorns that they have hidden.

The jays do well in this deal, as they loudly proclaim. But what about the oaks? It's very nice to have a predator eat the larvae that would otherwise eat your leaves, but not so nice when the predator seizes your fruit and eats it, or manhandles your seedlings. Yet the fact is that the jays are the world's great cultivators of oak and a principal tool in the oak's spread and dominance.

Jays carry acorns out of the deep shade of their adult trees, where they could never thrive, and plant them on forest edges, where the acorns have plenty of sun but some protection from wind. They plant the acorns in soils that are neither compacted nor waterlogged, since both resist a jay's effort to place the nut. The acorns have a good soil in which to grow, one that will not subject them to fungal infection or to asphyxiation. The jays cover and conceal the acorns, making them a less easy prey for the deer, mice, voles, turkeys, bears, porcupines, badgers, and others that love to feast on them.

On average, for every four acorns that the jay hides, it recovers only one. Even if many more are found by various other predators, or become infected and rot, or fall prey to jay foraging after they sprout, it is thought that at least a couple hundred out of the average jay's annual hoard of acorns survive to become viable seedlings. It doesn't take higher mathematics to see what this can mean for the propagation of oak forests.

Students of the postglacial landscapes of North America and Europe have long been puzzled by how fast the oaks recolonized the areas left bare by retreating glaciers. The ordinary route of spreading by wind pollination—however adaptable the oak genome—could not account for the one- to two-mile-per-year dispersal rates indicated by the distribution of fossil oak pollen. Neither could squirrels, who compete with the jays to collect and hoard acorns and so also help to propagate oaks, but who only take an acorn about one hundred feet from the source.

But a jay can go almost a full mile before it decides to bury its throatful of acorns. With the rain of light oak pollen following, it just might happen that oaks would appear—on the scale of plant migration—to be running at top speed northward. They have probably done this again and again, ever since oaks and jays left Laurasia.

So it is hard to admire those strong, stable oaks in the park or street or woods or chaparral without also admiring the jays. Those things that scream at you, dart overhead, and occasionally drop an acorn or something worse on your head give the oaks legs.

Flexibility

There are no deciduous pines, no evergreen maples. Nor are there any deciduous spruce, nor evergreen ash, no deciduous junipers, no evergreen lindens, no deciduous cedars, no evergreen dogwoods. But there are both deciduous and evergreen oaks.

Where I grew up, on the coast range of northern California, it snows once in fifty years or so. Most, but not all, of our oaks were evergreen. An unclothed oak was a dead oak, a very sad thing in my childhood. I assumed therefore that most oaks were eternally green, but the few fallen dead leaves interested me. They might appear on the ground, a dozen at a time, usually in autumn but at other times too. Most were from coast live oak *(Quercus agrifolia),* and they were cupped, waxy, stiff, and sharp-tipped. Usually they had turned a pale yellow, with the black spots of some fungus that had helped to kill them. If they fell tips down, they could scud along the ground in the wind like scooting insects. If they fell tips up, they pricked bare feet or surprised the unwary who sat down on a chair set beneath the oaks.

Now I live in the northeastern United States, where all of the oaks are deciduous. But here I notice the ones that refuse to shed their leaves. Most are pin oaks, young ones, but older ones and even other oak species sometimes retain their dry, dead, withered, stiff, tan-brown leaves. They are my winter companions. No one who has lived all his life in a place like coastal California can understand the variety of noises in winter woods where deciduous and evergreen trees are mixed. You can tell where you are by the distinct winter sounds.

Wind awakens the snowy woods. A grove of hemlock or spruce make a whishing sound in the wind, and snags and stems creak. Deciduous forests make deeper rising and falling sounds. If a tree has a crack in it, the rubbing parts may make sounds like gunshots, enough to make a walker start and turn.

The homeliest winter sounds, to my ears, are made by the dead, persistent leaves of the oaks. Not happy, but homely. They rattle and hiss, like a kid sliding on gravel or kicking it. I can look and see which tree is making the sounds. Each tree sounds different. I find I am proud of their persistence, and I always promise myself to notice when the leaves finally fall. (If they didn't, the new ones couldn't come.) But I have not caught them at it yet. I turn around one April morning, and there are the dangling flowers and the mouse-ear beginnings of the new leaves.

The evergreen oaks evolved for warm climates, whether wet or dry. Photosynthesis continues all year in these climates, so there is a continual need for water to climb the stems, carrying minerals and water to the sites of food and energy manufacture. The leaf surfaces tend to be waxy, preventing sudden water loss and wilting; the leaves are usually whole, not lobed, meaning that they both heat and cool more slowly; and the air resistance that impedes transpiration tends to be greater over these whole leaves than it is over thinner or lobed leaves. Often, the leaves may be slightly cupped, protecting the water-transmitting stomates on their undersides from drying winds and steadying the rate of water loss.

This outer difference has a corresponding inner difference. The evergreen oak, unlike its deciduous relations, does not show a pronounced shift between the large-diameter vessels of early wood, the wood produced at the beginning of the growing season, and the small, thick-walled vessels of late wood, the wood produced during the latter part of the growing season. The vessels that serve as the tree's circulation system are scattered more evenly through each annual ring.

Less water moves through these smaller pores, and it moves more slowly. Yet also it moves more surely. The smaller vessels are less likely to develop gas bubbles that break the water column, rendering that channel useless. Most of the evergreen's vessels function not only during the current year, but for several years after as well. The evergreen system creates a slow and steady pulse of life.

The deciduous oak represents the opposite choice. Though evergreen and deciduous oaks occur together in many climates, the deciduous comes into its own where the weather has a distinct cold season. When a deciduous oak's leaves first unfurl in the spring, they signal the cambium, the tree's growing layer, as far away as the root, and very quickly, a new ring of xylem cells, those which will conduct water from the ground to the leaves, blow up to more than five hundred times their original size. At the same time, xylem tissue starts expanding from each leaf bud's base downward, until the two meet and a new year's circulation system is inaugurated. Each xylem cell dies soon after it expands. Evacuated of a cell's usual internal contents, the xylem cells join end to end and in millions of combinations—vertically, horizontally, and spirally—to create a labyrinth of rising pipes.

The suction generated in an oak's xylem can be four times greater than the minimum needed to lift water from the root system to the top of the tree. The system makes a vacuum more perfect than any that can be made in a laboratory. When the sun is warm and water is available, more than four gallons of water per hour may pass through a single eight-inch diameter trunk. In high spring, the rising sap reaches velocities of two hunded feet per hour, or better than three feet per minute.

As the season progresses, though, most of these enormous sap superhighways suffer embolisms—bubbles of gas that block the water's flow. Sometimes this is caused by variation in pressure from day to day and season to season, but often it may be caused simply by too much tossing of the branches in the wind. When one vessel is blocked, of course, another path will be sought, but by summer, a large proportion of paths are likely blocked.

In a seeming anticipation of this, the deciduous oaks start making the much smaller, thick-walled vessels of late wood in late spring. After the spring binge, the deciduous trees become more like their cousins, the evergreens. Though the smaller vessels are slower to transmit the materials from the soil, they are surer to be passable, since their small size keeps pressures high and discourages bubble formation.

The size and shape of deciduous oak leaves helps to regulate their water economy. Slender leaves, like those of the southern live oak *(Quercus virginiana)* or of the aptly named willow oak *(Q. phellos),* experience less air resistance and so are able to rapidly give off water vapor. At the same time, however, their comparatively small size makes them quicker to cool, so that their rate of flow slows rapidly in the evening. The large, lobed leaves of red, black, or white oak *(Q. rubra, Q. alba,* and *Q. velutina)* experience more air resistance, but they heat up more and are capable of transpiring phenomenal amounts of water. But because the deeply dissected lobes expose more surface directly to the air than would unlobed leaves, these comparatively large leaves also cool quickly at evening.

If you had to characterize the difference between evergreen and deciduous oaks, you might say that the former were adapted to the marathon and the latter to wind sprints. And between these two pillars exists every possible oak intermediate. There are small-, whole-leaved deciduous oaks, and huge-, lobed-leaved evergreen oaks. There are oaks like the southern live oak or the pin oak *(Q. palustris),* whose leaves are not quite evergreen but are "persistent," remaining on the tree until the new leaves appear. There are oaks like the deciduous black oak, with immense, lobed leaves, whose pubescent hairs are thought to help cool the leaves' engines. There are oaks like the swamp white oak *(Q. bicolor),* which appear undecided as to whether to have lobes or not, and others like the post oak *(Q. stallata),* which appear to be trying to turn their leaves into small crosses. There are oaks like the English oak *(Q. robur),* with slender, shallow-lobed leaves that seem ready to wilt if you simply looked crossly at them, while the leaves of the bear oak *(Q. ilicifolia)* are so tough, spiny, and arch-backed they look like thousands of tiny reptiles about to attack.

One can't help thinking that there is an element of play in these large variations, but it must be adaptive, directed play. In

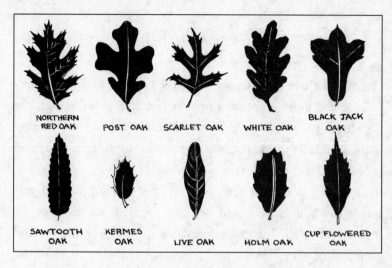

A few oak leaves (Nora H. Logan)

the very same habitat, there seem to be oaks that bet one way and oaks that bet the other. Generally, tree architecture tends to specialize either in outracing others for dominance in the forest canopy or in spreading fast to crowd the others out. The oak does both.

Up in the folded hills of the northern Sonoma wine country, I found very tall deciduous California white oaks *(Q. lobata)* living alongside wide-spreading evergreen canyon live oaks *(Q. chrysolepis)*. On one homestead on a Pomo Indian reservation, the tallest tree in the neighborhood—a white oak—lived across the street from the widest tree in the neighborhood—a live oak. The latter was so broad that the owner had built a fence around the drip line and made the enclosure into a hen coop and pigsty.

Prudence

In the tale of Sir Gawain and the Green Knight, the evergreen holly giant does not need saving. It is Gawain, the deciduous oak, who has to be saved from death. This is neither a quaint legend nor an archaic understanding. It tells truly and shortly the reality of leaves.

Leaves are not decorative. Not for trees and not for us. Little but microorganisms would be alive on earth were it not for leaves. When the leaves of deciduous trees appear and expand each spring, they signal a new season of harvesting energy from the sun. None but they can do it. Without them, we all die.

The sprouting of the new leaves is called "flushing" or a "flush." Some trees have just one flush of new leaves, in the spring. Others manage two flushes. Some add leaves continuously from spring through fall.

All of these methods have advantages and disadvantages. One flush may be fine if spring is dependable, but bad weather just at that time may result in many damaged leaves and so seriously reduce the plant's ability to make food for the rest of the year. Two flushes provide some insurance against bad weather, but bad luck is still too likely. You would think that continuous growth would solve the problem, but the trouble is that it costs a great deal of energy to keep pumping out new growth. If bad weather is concentrated in one season, repeated efforts to send out leaves can seriously reduce the tree's energy reserve.

The deciduous oaks have evolved a compromise that takes advantage of the energy economy of discrete flushes and the long time frame of continuous growth. The deciduous oak may flush up to four times: in early spring, early summer, high summer, and sometimes one in the fall. In other words, an oak may sprout up to four times from its buds each year. The discrete flushes do not depend on the success of the previous flush, but the tree rests between flushes so that it does not waste all its energy against a climatic anomaly. In this way, even in a bad-weather year, the oaks have a chance at growth and new life. And if conditions change for a number of years in a row, they can adjust the number of flushes to fit the new climate.

At each flush, a new set of tip and lateral buds is formed in the terminal bud of the previous flush. When the flush hits, the terminal buds open and leaf out at the expense of the lateral buds, which usually remain dormant. This is why oaks may seem to change color in midsummer: The suddenly expanding new growth may be reddish or yellowish in color, and because it is held out stiffly on the terminals, it seems to be a whole fresh set of leaves covering the trees. The laterals behind offer no competition.

What is happening behind this bright show, however, accounts for a lot of the oak's adaptability and its wide, space-filling growth. The strongest lateral buds of the previous flush—which had done nothing when their tip buds were expanding—now sprout and expand. Behind the out-thrust scouts of the new terminals, then, phalanxes of lateral twigs appear all around the branch. And what's more, these laterals may outgrow the terminals, the main shoots. So behind the up and outward thrust of each succeeding flush, the oak completely occupies the recently conquered space.

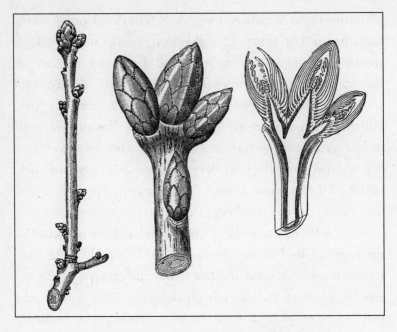

Oak buds (from H. Marshall Ward, *The Oak*)

The oak is profligate not only in the number of lateral buds that grow, but also in the number of lateral buds that do not. Typically, only the outermost and strongest laterals grow, leaving more than half of them dormant. This means that in the average mature oak, perhaps a million buds will never sprout at all. Liberty Hyde Bailey, the American horticulturist whose fine observations and careful illustrations of growing stems alone are worth the price of his books, thought this a fine demonstration of the fact that nature was red in tooth and claw. The buds, he said, compete for the right to grow, but only the strongest and fittest actually get to.

But this is not exactly so. The dormant buds are not the losers; they are part of the reserves. They may remain viable for more than a year, and if necessary—say, after the loss of part of a branch in a storm—they sprout. And where dormant buds fall, the tree can produce new buds called adventitious buds, from scratch.

This reserve was crucial not only to the oak's survival, but to the health of human communities. The wooden world depended on oak for everything from firewood to structural timber, and much of it was obtained by cutting the tree back—either to major branches or right to the ground. You could count on the oak to regenerate strongly from these cut ends and create fine timber within a decade.

Dormant and adventitious buds are also part of the oak's persistence and its flexibility. Humans are not the only agents that damage or destroy branches. There are also caterpillars, beetles, twig girdlers, borers, cankers, rots, thunderstorms, ice loads, hurricanes, goats, cattle, and sheep. And who knows what else feasted on the tender leaves and twigs of oaks in pre-human times? The buds are the oak's answer to destruction. They are the sites from which, no matter what, growth can be renewed.

They are also a gift to its decline. No tree grows old so beautifully as the oak. If you could compress nine hundred years of time into a few minutes, you would see an oak as a firework, shooting up, expanding, and then like the best rockets, seeming to sink back into itself while a rain of green drops all around.

In its sinking, the oak becomes stag-headed, its dying leaders sticking out like the horns of a buck. The branch ends begin to die and to fall away. With every death, a few new buds sprout and keep up the show, together with the lower branches. The sprouts often go straight up and then curl over like water in a fountain. Sometimes there are many on a branch, sometimes few. The oak does not go quickly and quietly. It celebrates its own demise. This is why the poet John Dryden said of it, "three hundred years growing, three hundred years living, three hundred years dying."

Persistence

The ratio of roots to shoots in a plant is a standard measure among botanists and a rough way to express the staying power of a young tree. If a tree puts on a lot of top growth and few roots, it is liable to be weak wooded and short lived. Fast growers mature young and reproduce in haste. The birch, for example, dwarfs the oak in its capacity to produce seed, making orders of magnitude more seeds in a life that it is only a quarter as long. The early death of a few thousand individuals is of little consequence to such a tree. If a tree puts down a great deal of roots and adds shoots more slowly, however, it is liable to be long lived and more resistant to stress and strain.

Few seedling trees have a root-to-shoot ratio that is greater than one, and in most trees it is less than one. This means that in most trees there is more shoot mass aboveground than there is root mass belowground. But the average ratio for a seedling oak is between four and six, and there have been oaks whose ratio was greater than ten. A seedling oak may have ten times as much root mass as it does stem and branch mass.

When the acorn first ruptures, surface dwellers often don't even know about it. Likely, the lucky acorn is hidden beneath a layer of leaf and litter. The future trunk has not yet lifted above the surface. But the root is already heading down fast.

Oaks make lots and lots of roots. No matter what, they stand a chance of survival. The British botanist M. W. Shaw reported finding oak seedlings less than two feet tall that were twenty-five years old. Within the first year, the taproot of an oak can reach more than a foot in depth, thickening to about the width

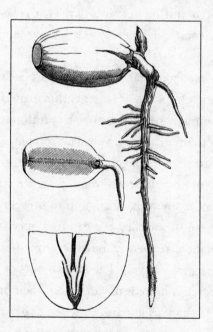

The root going down (from
H. Marshall Ward, *The Oak*)

of a pencil and sending out its first lateral roots all along the stem. Though the taproot later declines in importance, it spends the first few years thickening and storing as much food as it can, for until the oak's canopy is well above the reach of everything from turkey to deer, it is liable to be chewed down again and again.

Some root tips circumnutate, that is, they "swim around." The meristem cells—that is, the growing tips—divide, then stretch, then divide again. The fact that they do this at varying rates according to their position around the circumference of the root accounts for the twisting motion. The root is like an auger, opening the way in the soil. Roots can travel pretty fast, up to twenty-five millimeters in a day. That is fast enough that were you patient (and able to see through soil), you could watch them grow.

Half the total food that the tree makes each year goes to grow new roots. One student laments this inefficiency and is at a loss to explain why oaks should be so profligate. I think that the oak is practicing, the same way a person plays the same succession of musical notes again and again to know them by heart. By putting out far more roots than are strictly needed, the oak is more likely to have enough to prosper.

Roots may spin, but they try to grow outward in straight lines. They branch frequently, but the growing tip of any one root keeps going as it began. The lateral roots of a large, open-grown oak stretch hundreds of yards from their origin. At first, they may be as thick as a man's torso, but they quickly dwindle until six feet from the trunk they are only as thick as his wrist, at thirty feet they are as slender as a pencil lead, and at fifty feet as thin as grass, and at one hundred feet thinner than a human hair.

Roots do not simply swim. They sense and respond to irritants—rocks, water-saturated zones, poisons—as finely as the human tongue, and they are stronger and more intelligent than a human finger or arm. If a growing root encounters a stone, it may compress slightly, deflect to one side, and return to its straight line once the obstacle has been passed. Or, if it senses

A root that squeezed itself
flat to pass through the crack
in a rock (from H. Marshall
Ward, *The Oak*)

the slenderest crack in the rock, it may splay out thin and grow right through the fissure to the other side.

Roots know where they are. They know how deep they are in the soil, because they measure the amount of oxygen around them. Where there is more oxygen, there is more growth. This is why the cambium—the layer of cells just below the bark that is responsible for the thickening of stems—usually puts on more new tissue on the upper side of a root growing parallel to the ground than on its lower side. If a root has to twist out of the way of a stone, it has the means and knowledge to return from the deviation, since it will seek to reestablish its preferred oxygen environment.

It was once thought that the tree's branches and its roots were mirrors of each other. If you took the tree out of the ground and held it by the middle of the trunk, you would have something that resembled a hairy dumbbell. But this is not so. Botanists found this out when some extremely patient Scandinavians painstakingly, with the care of great painters or archaeologists, scraped away the soil from the root systems of several mature trees.

Imagine their excitement and exasperation. They had thought they would run out of roots at about the same distance from the bole as they ran out of branches overhead. No, the roots kept going. Five feet, ten feet, twenty-five, fifty . . . As the roots got smaller and more numerous toward the tips, the scientists must have been giddy with fatigue, eyestrain, and the sense of discovering something very important while doing something virtually impossible and completely ridiculous. They gave up when the distance of the root mass from the trunk was nearly three times as wide as the maximum spread of the tree's crown. And there were still more roots beyond.

The smaller the roots get, the shorter they live. Most smaller roots live four years or less, some for as little as a season or even a week. Many of the pencil-lead and thinner roots die and are replaced several times in one year. Root hairs—the almost microscopic hairs that work with beneficial fungi to bring nutrients into the tree—may be replaced every few hours. Yet the average mature red oak has better than five hundred million living root tips.

Root growth in oaks is doubly persistent. While the leaves sprout three or four times each year, the roots grow continuously from March to October, and in some climates into November. The biggest spike of growth occurs between May and late June in the temperate Northern Hemisphere, with a small spike between late August and early October.

It is not a good thing to risk damaging the roots, for it is these upon which the oak's life depends. Leaves and branches are the beauty of the tree, and one tends to think that they must be the most important part. Not so. You can cut the oak back to the ground again and again, and it will resprout from the energy contained in the roots. But once you seriously damage the roots there is a good chance that the tree will die.

Unseen, a cloud of hundreds of millions of root tips swim through the soil. The mystery writer Wilkie Collins was not just waxing poetic when he wrote, "Fancy and imagination, grace and beauty, all those qualities which are to the work of art what scent and beauty are to the flower, can only grow towards heaven by taking root in the earth."

Community

Many individual oaks aren't. The live oaks of central Texas, for example, for all that they form groves stretching over fifty square miles, are really each composed of a small number of individual oaks that have made many auxiliary trees by sprouting from their own roots. In other words, a grove that appears to comprise hundreds of trees may really only consist of four or five genetic individuals.

But even where oaks are born as individuals, they often do not remain so. As an oak matures, its roots reach an area four to seven times as wide as the width of the tree's crown, the diameter of their spread about twice as wide as the tree is tall. This means that in a mixed forest dominated by oak the soil is densely packed with winding roots. Roots of different species of oak, or of different genera, may wind tightly about each other and may hold their fans and mats of infinitesimal, mycorrhizally infected feeder roots cheek by jowl in the same parcels of decaying leaf litter. Nevertheless, they remain independent of each other. But within the same species of oak—particularly for species in the red oak group—when roots meet, they very often graft, that is, their vascular systems join and in essence, they become one flesh.

The old saw "You can't see the forest for the trees" is truer than metaphor. Because we cannot see beneath the soil, we do not know that these oaks are not simply neighbors, providing mutual shade, protection from the wind, and sharing predator pressures, they are physically united beneath the ground.

In any forest, some trees are dominant. They reach the light first, spread fastest, occupy territory both above- and below-ground. Other trees—foresters call them "suppressed"—have trouble reaching the canopy at all and their trunks may be only half the diameter of the dominant trees, though they are the same age. Still others may have started much later and so have little chance to reach the crown. Other species, too, may grow faster than young oaks and so deprive them of light.

In a world where nature is said always to be red in tooth and claw, one might well expect this to be the norm: If you can't compete, then just die and get out of the way. The fact is, however, that the dominant oaks in an oak forest sometimes support the suppressed ones. Food travels through the grafted roots from the trees with extra food to those that do not have enough. The forest supports its weaker members. This turns out to be a good thing for the entity as a whole, since when the largest tree succumbs to a storm, to old age, or to axes and saws, the suppressed trees may be healthy enough to take over the dominant position.

The forest supports its sick trees too. Say an oak is suffering from a partial girdling, and nutrition cannot flow down from the photosynthesizing leaves to the base of the plant and to the roots. Instead of dying of starvation, the tree is fed by sugars that are imported from other, healthy oaks through the root grafts. The borrowed sugars ascend the stem in the healthy xylem and buy time so that the wounded tree can rebuild its own circulation.

In fact, the forest even supports its dead. Stumps in an oak forest are very slow to decay, and they will often sprout again.

This is not simply because they are "vigorous" in themselves, but because their roots may be supplied with nutrients by the neighbor roots to which they are grafted. If you look closely at some forests of pin or red oak in America's upper Midwest, you will see that the majority of the trees are stump sprouts, regenerated from a forest logged decades ago.

Even where a stump does not regenerate, its roots may go on serving the forest. The grafted roots may go on acquiring water and nutrients for the surviving trees long after their parent tree has rotted away. These "adopted roots" are common in oak forests.

All this mutual help, however, comes with one grave disadvantage. The roots are an open pipeline for food, but they might also be a conduit for disease to spread quickly and without barrier. If the forest were merely a community, there might be hope: The physically separated trees might be spared, just as the storytellers of Boccaccio's *Decameron* hoped to be spared from the plague by locking themselves up in a castle and avoiding all contact with the diseased. But where the members are connected underground, there is no hope of isolation.

Generosity

The list of guests at the great oak feast is longer than the Catalogue of Ships in the *Iliad*. Among the creatures that feed at the table are the great prominent, the oak hook tip, the thorn moth, the oak tree pug, the brindled and marbled pugs, the oak beauty, the great oak beauty, the green oak tortrix, the

blotched emerald, the purple emperor, the wood worm, the oak
bark beetle, the red oak roller, the nut weevil, the leafhopper,
the buzzing and jumping spiders, the sawflies, the deathwatch
beetle, the tent caterpillar, and the tussock moth, the oakworm
and the oak twig pruner, the two-lined chestnut borer and the
Asiatic oak weevil, the pigeon tremex, the periodical cicada,
the greedy and golden-oak scales, the obscure scale, the oak
leucanium, the lactarius, the boletes, the russellas, the armil-
larias and the ceratocistus, the truffles and the ganodermas, the
mice and the voles, the shrews and the badger, the squirrels,
the turkey, the wood pigeon, the fallow deer, the mule deer, the
red deer, the white-tailed deer, the porcupine . . . There are
many more. But none is so intimately related to the oak as the
gall-making wasps, the cynipines.

"The oak," wrote Theophrastus in about 300 B.C. in his
Enquiry into Plants, "bears more things besides its fruit than
any other tree." He noticed a dozen different galls: one a scarlet berry; another a "black resinous gall"; one like a phallus;
another like a bull's head; one soft, woolly, and spherical, which
could be used as a fire starter; another a hairy ball that in
spring sweated sweet juice. He recorded spot galls and hollow
galls, button galls and horned galls, red, yellow, black, and
transparent galls. Not only that, he said, the oak bore many
fungi that grew from its roots. And he cited Hesiod as claiming that both honey and bees were also produced by the oak
tree, though being perhaps a better observer than the poet,
Theophrastus noted that Hesiod may have mistaken the honey-
sweet droplets often found on oak leaves—the aphid drippings,
as we know now—for the precursor and origin of honey.

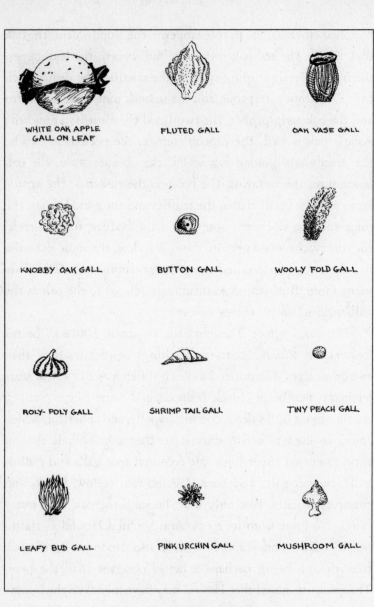

WHITE OAK APPLE
GALL ON LEAF

FLUTED GALL

OAK VASE GALL

KNOBBY OAK GALL

BUTTON GALL

WOOLY FOLD GALL

ROLY-POLY GALL

SHRIMP TAIL GALL

TINY PEACH GALL

LEAFY BUD GALL

PINK URCHIN GALL

MUSHROOM GALL

A few oak galls (Nora H. Logan)

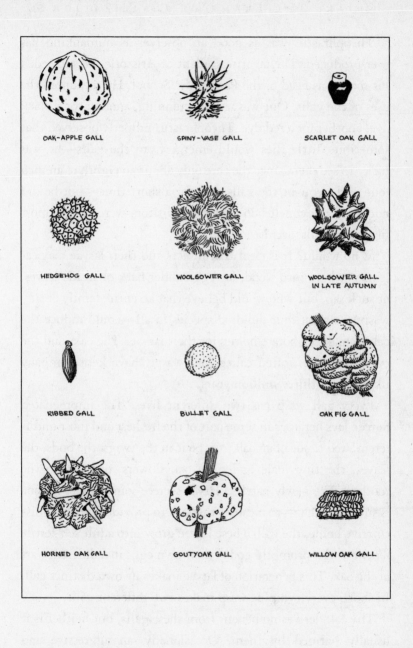

OAK-APPLE GALL   SPINY-VASE GALL   SCARLET OAK GALL

HEDGEHOG GALL   WOOLSOWER GALL   WOOLSOWER GALL IN LATE AUTUMN

RIBBED GALL   BULLET GALL   OAK FIG GALL

HORNED OAK GALL   GOUTYOAK GALL   WILLOW OAK GALL

Theophrastus was as good an observer as humankind has ever produced. The favorite student of Aristotle, he succeeded his master as head of the Peripatetic School. He recounted the uses of the galls. One was used for tanning, another for a black dye, another for a red dye. Theophrastus properly observed that sometimes little flies would emerge from the galls—he was lucky to see them since they are only about an eighth of an inch long and frequent the galls only for a short time—but he did not draw the conclusion that these visitors were in fact most likely the galls' architects.

Who would? It is clear that insects and their larvae may eat leaves or bore into wood, lay eggs under bark or inside leaves, or suck sap, but who would believe that an entire family of tiny wasps—almost four hundred species in all—could induce the oak to make elaborate homes for their larvae? The oaks and the cynipine wasps, often called gall wasps, have kept company like this for thirty million years.

Every gall wasp has two different lives. The impregnated female lays her eggs in some part of the freshest and most undifferentiated tissue of an oak tree, be it in the twigs, the buds, the leaves, the tiny male or female oak flowers, or even on the rootlets. The newly hatched larvae secrete something—no one is sure just what—that causes that oak to provide it with its distinctive home, the gall. These larvae grow into adult self-fertile females, who promptly go lay their own eggs in a different part of the oak. This generation of larvae makes its own distinct gall. In short, every cynipine wasp makes two different galls.

The oak derives no benefit from these galls, but neither is it usually harmed by them. Occasionally, an aggressive and numerous maker of twig galls—like the wasps that make the so-called horned and gouty oak galls—create so many galls that branches are girdled and die. (Such oaks appear to have sprouted black golf balls.) But this is the exception and only occurs to oaks under stress for other reasons. Mainly, the oak acts as the bricks and mortar that are deployed according to the larvae's plans.

The cynipine galls are extraordinary, many-layered and labyrinthine. They exist not simply to feed growing larvae but also to defend it against the weather and especially against the numerous parasitoids that would either eat the larvae or take over the house for their own broods. Every species of cynipine has up to twenty other species of arthropods and countless species of fungi that, given the chance, would devour its larvae and move into the vacated space.

A gall is a castle. Usually, the inner layer is a rich sphere dripping with food that the larvae graze upon with their chewing mouthparts. It may contain one or numerous larvae, depending upon the species. Surrounding this sphere is usually a hardened layer, resistant to the piercing mouthparts or egg-laying tubes of parasitoid mothers. Outside this is a spongy layer very rich in tannins and in other acids that the gall's tissue somehow concentrates to as much as ten times their level in normal oak tissue.

Still, the contest between larvae and would-be predators is a serious one. Some of the invaders can even digest tannins, and in fact hollow out a little space in the tannic layer where their own young are born and grow, sharing quarters with the gall wasp larvae. The cynipines are not new to this situation. They have had thirty million years to prepare and elaborate further defenses in the outer layers of their gall homes. One solution is to make the shape a puzzle: The pink to purple gall of one wasp looks rather like a sea urchin, but the larval chamber is at the base of only one of the many spikes. The warty and horny twig galls of another are a labyrinth of larval chambers, only a few of which contain the living treasure.

A second solution is to outrun your enemy's firepower. The chalcid wasp's weapon against the cynipine is the ovipositor: a penetrating lance that holds on its tip an egg that is deposited at the uttermost limit of its extension. The cynipines, in response, engineer a mature gall just a little bit thicker than the ovipositor is long, so the frustrated chalcid is left with its egg just shy of the juicy larval chamber. Many galls, upon examination, turn out to have ovipositors broken off in them. The predator's only chance is to attack when the gall is still young and the diameter of the protective tissue is less.

A third solution is to cover the gall's surface with something unappealing to attackers. Some galls are quite sticky and threaten to retain the would-be mother like a fly on flypaper. Others are hairy and difficult to alight upon at all, or spiny and hard.

But perhaps the most novel and effective idea is to hire protection. Ants will furiously attack anything that threatens their herds. Ants, as is well known, farm aphids, scales, and other invertebrates, milking these creatures for the sweet honeydew that they draw from the plant's sap. Certain galls themselves have developed a means of synthesizing and secreting a similar honeydew, so that ants gather around the galls, feed upon the honeydew, and fiercely defend the galls.

The average oak, then, is like a country dotted with gall towns. Each town has its weather, its streets and walls, its inhabitants, its attackers, and a life span. Once the cynipines move out—leaving holes in the walls of their homes—fungi immediately enter and begin to digest the walls. I have a collection of old galls—big hollow balls, horned tunnels, red spots, odd clown's hats—and each of them reminds me of ruins. To feel one and turn it over in my hand is like visiting an abandoned village.

EIFFEL AND OAK

THEY BUILT THE Eiffel Tower for the Paris Exposition of 1889. At precisely three hundred meters tall, it was almost twice as tall as the next largest monolith, the Washington Monument. It took more than 2.5 million rivets to hold together its 18,038 pieces of wrought iron, and the thing weighs more than 7,000 tons. Nevertheless, it took only a little more than two years to build. George Berger, general manager of the fair, declared the meaning of the achievement: "We will give [the people] a view from the steep summit of the slope that has been climbed since the Dark Ages. . . . For the law of progress is inexorable, just as progress itself is infinite."

The Eiffel Tower is the founding monument of industrial modernity. It had and has no purpose but to astound. The official Eiffel Tower Web site gives among its statistics the following: "Distinctive Feature: Recognizable throughout the entire world." In short, it is an enormous trademark.

At the start, not everyone was mightily pleased with the new view. Before the tower was half completed, a group of Paris artists that included Paul Verlaine, G. K. Huysmans, and Guy de Maupassant wrote a letter of protest that appeared on the front page of *Le Temps,* the chief Paris daily. The artists compared the tower variously to a skeleton, a factory chimney, a gymnast's apparatus, and an immense suppository.

Alexandre-Gustav Eiffel responded in *Le Temps* that his structure would be seen as beautiful, because it conformed to the "hidden rules of harmony." He added, "Moreover, there is an attraction in the colossal and a singular delight to which ordinary theories of art are scarcely applicable." It was better *because* it was bigger.

The French are expert at insults. One of the best hurled against the tower came from a fierce writer named Leon Bloy. Looking on Eiffel's Tower, he was moved to dub it a "truly tragic street lamp."

The official Eiffel Tower Web site itself replays this "controversy" in the manner of all good press agents. (Any press, good or bad, is good for business.) And it resolves the dilemma in the accepted way: Two million people visited the tower during the Paris Exposition, and more than 180 million more have visited it since. The people have spoken: The tower is beautiful because it has had so many visitors.

But if you will call to mind your own experience in visiting this or any other spectacle of size—the Seattle Space Needle, or the Empire State Building, or the lamented World Trade Center—you may recognize a kind of dutifulness in the visits. You wait forever for the elevator, you crowd into it, and you are disgorged on the summit. After a moment of vertiginous shock, you look around. Dutifully, you continue to look around. You try to recognize monuments dwarfed in the distance. There is a palpitating emptiness to the exercise, as great I don't doubt as the emptiness experienced by nineteenth-century men and women who entered the cathedrals only because it was what everybody had to do.

They were supposed to believe in God; we are supposed to believe in progress. But this summit we have reached in these monuments is somehow arid. Perhaps God is not what they had been taught by their priests, and perhaps progress is not what we have been taught by ours. C. S. Lewis wrote that when you are on the wrong road, the shortest way to go forward is to go back to where you made the wrong turn and make the right one.

Compare the structure of an oak tree with the Eiffel Tower. There is an obvious analogy. It was noted by D'Arcy Thompson in *On Growth and Form*. Both oak and tower are basically long cones that flare at the base. Eiffel was proud of his tower's basal flare. His mathematical calculations, he said, indicated the dimensions necessary at the base so that the tower could stand up to the wind.

The oak too has a pronounced flare at the base, more so than any other tree. If you are walking in a dense forest where all the leaves are high overhead, in fact, you can tell which are oaks by looking for the basal flare. The reason is the same, for though the oak is shorter than the tower and made of lighter materials, it is solid where the tower is skeletal, and its crown catches the wind.

But why does the oak have such a pronounced flare when other tall trees do not? A fast-growing tree like poplar or ailanthus has weaker wood than the oak. The fast grower puts its energy into rapid expansion, not into firm structure. Therefore, when a wind- or snowstorm comes, these quick trees respond by shedding branches or even by losing their crowns. The oak seldom does. When an oak fails, it most often fails from the base, usually when the ground gives out. The roots break and the whole tree topples at once. The basal flare is wood added to strengthen the roots at their base and so reduce the chances of failure.

The second analogy between the Eiffel Tower and an oak is in their internal structures. Though Eiffel may have noticed the flare at an oak's base, it is unlikely that he ever knew how much the skeletal structure of the tower is a parody of the structure of oak tissue. Two things kept solid stone structures like the Washington Monument or the Gothic cathedrals from rising even higher: their weight and their resistance to the wind. Eiffel used a skeleton of iron—a material stronger than stone but no heavier—so that he could construct a much taller object without creating unbearable forces of compression. Vertical members are reinforced by diagonal members and stabilized by cross members. All the parts are joined by rivets, and there is more open space than iron in the volume of the cone. The wind passes freely through the structure, and the jointed frame is even able to adjust slightly to let it through. "Eiffel was one of the first," wrote Maurice Besset, "to create a form not as neutral and stable in space, but living and moving." Yes, if you discount the sixty-five-million-year history of the oaks.

To all appearances, the tower is just the opposite of the oak, which is quite solid and full of large leafy branches. Far from letting wind pass freely through, the oak tends to catch it and to shake its head furiously in the storms. But the reason it can do this is an internal structure that makes the tower's look rudimentary. The vertical fibers that make up oak flesh are crossed by rays of living cells that stretch inward from the bark to the pith, stabilizing the framework and keeping the vertical fibers from splitting. Furthermore, the living stresses exerted by the tree's own growth—the swelling of the layers of living tissues near the trunk surface—push the fibers together into a tight, hard-to-break bundle. Third, the fibers are woven through each other, but they are also discrete pieces, so that they can flex and slide over one another without shearing or destroying their vital connection. No rivets are needed.

Klaus Mattheck, in his continuing study of tree structure, has explained wood structure brilliantly. The tree, he says, has three kinds of structural members: bricks, ropes, and I beams. The bricks are the lignin—strong hydrocarbons that form the framework of every cell in the tree's fibers. They resist the compression that comes from the weight of the tree, and the compression on some fibers when the tree is bent under wind or snow. The ropes are the cellulose, the core of every fiber cell, which are bendable and very hard to break. These, he writes, act to resist tension, for when the tree is bent, some of the cells are necessarily compressed while others are stretched. The I beams are the ray cells—bands of live tissue that stretch from bark to heart. These prevent the fibers from shearing and parting, even when the wind puts twisting loads on the stems.

Indeed, it is only thanks to this internal structure that a tree can be solid instead of skeletal. If you put a skin on the Eiffel Tower or gave it branches, the thing would fall down in the first good blow.

Iron is harder than wood. This is perhaps the lone advantage of the tower's structure. But this advantage is counteracted by the fact that wood, though softer, is dynamic, while iron is passive. The Eiffel Tower is one large skeletal iron cone. The mature oak is many cones, each sheathed one on top of the other by year after year of annual growth. It is enjoyable to envision the tree—not only the trunk but the roots and branches—as hundreds and hundreds of cones. The outermost cones contain the living cells while the inner cones are filled with plugs and tannins and compounds that further stiffen the structure. Each year, a new set of cones is added, the pressure of growth helping to hold the stems together while the inner wood stiffens to resist both physical and biological stresses.

The oak is generative, the tower parasitic. To renew its outer protection, the oak has a cork cambium that annually generates an increment of waxy skin to protect the tree from damage. In fact, it is the cambium's old layers, flaking off as they are squeezed outward by the expanding trunk, that make the rough texture and pattern of the bark that we feel. The Eiffel Tower, instead, needs outside help every seven years to keep it from rusting away. To protect the iron from the weather requires fifty tons of paint and a full year just to apply it.

What do the two structures provide in exchange for the energy expended to maintain them? The oak gives oxygen to the air. More than five thousand different species live on, in, or by means of the average oak. Even today, oak is used by man for furniture, packing pallets, railroad ties, flooring, wainscoting, timber frames, basketry, porridge, firewood, and charcoal. On a summer's day, it is up to ten degrees cooler in the shade of a great oak.

What does the tower give? A spectacle. To see and to see from. And on a hot summer's day, you practically fry standing on its treeless plaza.

If you had to pick one or the other to emulate, which would you choose?

BIBLIOGRAPHY

Note: **Boldfaced** entries are key resources.

GENERAL WORKS

Bechmann, Roland. *Trees and Man: The Forest in the Middle Ages.* New York: Paragon House, 1990.

Coutance, A. *Histoire de chêne dans l'antiquite & dans la nature.* Paris: J.B. Bailiere et fils, 1873.

Edlin, H. L. *Trees, Woods and Man.* London: Collins, 1956.

Frazer, James George. *The Golden Bough.* London: Oxford University Press, 1994.

Graves, Robert. *The White Goddess.* New York: Farrar, Straus and Giroux, 1948.

Heinrich, Bernd. *The Trees in My Forest.* New York: HarperCollins, 1997.

Jackson, James P. *The Biography of a Tree.* Middle Village, N.Y.: Jonathan David, 1979.

Keator, Glenn. *The Life of an Oak: An Intimate Portrait.* Berkeley, Calif.: Heyday Books, 1998.

Loudon, J. C. *Arboretum et Fruticetum Britannicum, or, The Trees and Shrubs of Britain*. London: Longman, Orme, Brown, Green, and Longmans, 1838. Vol. III, pp. 1717–1931.

Meiggs, Russell. *Trees and Timber in the Ancient Mediterranean World*. Oxford: Oxford University Press, 1982.

Morris, M. G., and F. H. Perring. *The British Oak: Its History and Natural History*. Berkshire, UK: The Botanical Society of the British Isles, 1974.

Mosley, Charles. *The Oak: Its Natural History, Antiquity and Folklore*. London: Elliot Stock, 1910.

Peterken, George F. *Natural Woodland: Ecology and Conservation in Northern Temperate Regions*. Cambridge: Cambridge University Press, 1996.

Rackham, Oliver. *Ancient Woodland: Its History, Vegetation and Uses in England*. London: Edward Arnold, 1980.

———. *Trees and Woodland in the British Landscape*. London: J. M. Dent, 1976.

Tansley, A. G. *Oaks and Oak Woods*. London: Methuen, 1952.

Thoreau, Henry D. *The Journal of Henry D. Thoreau*. Edited by Bradford Torrey and Francis H. Allen. Vol. 14 (August 1, 1860–November 3, 1861).

Ward, H. Marshall. *The Oak: A Popular Introduction to Forest-Botany*. New York: D. Appleton and Company, 1892.

Williamson, John. *The Oak King, the Holly King and the Unicorn*. New York: Harper & Row, 1980.

BALANOCULTURE

Bainbridge, David A. "The Rise of Agriculture: A New Perspective." *Ambio* 14, no. 3 (1985): 148–51.

———. "The Use of Acorns for Food in California: Past, Present, Future." Berkeley, Calif.: U.S. Department of Agriculture, Pacific Southwest Forest and Range Experiment Station, Gen. Tech Rep. PSW-100 (1987).

Barfield, Lawrence. *Northern Italy Before Rome*. London: Thames and Hudson, 1971.

Baumhoff, Martin A. "The Carrying Capacity of Hunter-Gatherers." In *Affluent Foragers: Pacific Coasts East and West*, edited by Shizuo Koyama and David Hurst Thomas. Osaka: National Museum of Ethnology, 1979.

—————. *Ecological Determinants of Aboriginal California Populations*. Berkeley: University of California Press, 1963.

Bean, Lowell John. *Mukat's People: The Cahuilla Indians of Southern California*. Berkeley: University of California Press, 1972.

Bean, Lowell John, and Katherine Siva Saubel. *Temalpakh: Cahuilla Indian Knowledge and Use of Plants*. Morongo Indian Reservation: Malki Museum Press, 1972.

Bohrer, Vorsila L. "On the Relation of Harvest Methods to Early Agriculture in the Near East." *Economic Botany* 26, no. 2 (1972): 145–55.

Brouk, B. *Plants Consumed by Man*. London: Academic Press, 1975.

Chestnut, V. K. *Plants Used by the Indians of Mendocino County, California*. Fort Bragg, Calif.: Mendocino County Historical Society, 1974, pp. 333–44.

DeBois, Cora. *Wintu Ethnography*. Berkeley, Calif.: University of California Press, 1935.

Driver, Harold E. "The Acorn in North American Indian Diet." *Proceedings of the Indiana Academy of Science* 62 (1952): 56–62.

Fernald, Merritt Lyndon, and Alfred Charles Kinsey. *Edible Wild Plants*. New York: Harper & Brothers, 1943.

Flannery, Kent V. "The Ecology of Early Food Production in Mesopotamia." *Science* 147, no. 3663 (1965): 1247–56.

—————. "Origins and Ecological Effects of Early Domestication in Iran and the Near East." In *The Domestication and Exploitation of Plants and Animals*, edited by Peter J. Ucko and G. W. Dimbleby. London: Gerald Duckworth & Co., 1969.

Gifford, E. W. "California Balanophagy." In *Essays in Anthropology Presented to A. L. Kroeber in Celebration of His Sixtieth Birthday*. Freeport, N.Y.: Books for Libraries Press, 1936.

Goldschmidt, Walter. *Nomlaki Ethnography*. Berkeley: University of California Press, 1951.

Gravas, Robert. *The White Goddess*. New York: Farrar, Straus and Giroux, 1948.

Harlan, Jack R. "Self-Perception and the Origins of Agriculture." In *Plants and Society*, edited by M. S. Swaminathan and S. L. Kochhar. London: Macmillan, 1989, pp. 5–32.

Harris, David R., and Gordon C. Hillman. *Foraging and Farming: The Evolution of Plant Exploitation*. London: Unwin Hyman, 1989.

Heizer, Robert F., and Albert B. Elsasser. *The Natural World of the California Indians*. Berkeley: University of California Press, 1980.

Hesiod, in *The Homeric Hymns and Homerica*. Translated by Hugh G. Evelyn-White. Cambridge, Mass.: Harvard University Press, 1914.

Howes, F. N. *Nuts: Their Production and Everyday Uses*. London: Faber and Faber, 1948.

Jorgensen, Grethe. "Acorns as a Food Source in the Later Stone Age." *Acta Archaeologica* 48 (1977): 233–38.

Kidder, Tristram R., and Gayle J. Fritz. "Subsistence and Social Change in the Lower Mississippi Valley: The Reno Brake and Osceola Sites, Louisiana." *Journal of Field Archaeology* 20 (1993): 281–97.

Kroeber, A. L. *Anthropology*. New York: Harcourt Brace, 1923.

Kroeber, Theodora. *Ishi in Two Worlds*. Berkeley: University of California Press, 1961.

Lowenfeld, Claire. *Britain's Wild Larder: Nuts*. London: Faber and Faber, 1965.

Margolin, Malcolm, ed. *The Way We Lived: California Indian Stories, Songs & Reminiscences*. Berkeley, Calif.: Heyday Books, 1981.

Mason, Sarah R. "Acorns in Human Subsistence." Dissertation, University College, London, 1992.

Meltzer, David L., and Bruce D. Smith. "Paleoindian and Early Archaic Subsistence Strategies in Eastern North America." In *Foraging, Collecting and Harvesting: Archaic Period Subsistence and Settlement in the Eastern Woodlands*, edited by Sarah W. Neusius. Carbondale, Ill.: Southern Illinois University Press, 1986.

Merriam, C. Hart. "The Acorn, a Possibly Neglected Source of Food." *National Geographic* 34, no. 2 (1918): 129–37.

Muir, John. *My First Summer in the Sierra*. Boston: Houghton Mifflin, 1979.

The New American Bible. Nashville: Thomas Nelson Publishers, 1987.

Ocean, Suellen. *Acorns and Eat 'Em: A How-To Vegetarian Cookbook*. Potter Valley, Calif.: Old Oak Printing, 1993.

Opler, Morris Edward. *An Apache Life-Way*. Chicago: University of Chicago Press, 1991.

Ortiz, Beverly R. *It Will Live Forever: Traditional Yosemite Indian Acorn Preparation*. Berkeley, Calif.: Heyday Books, 1991.

Ovid. *Fasti*. Translated by Sir James G. Frazer. Cambridge, Mass.: Harvard University Press, 1931.

Pausanias. *Description of Greece*. Cambridge, Mass.: Harvard University Press, 1977.

Pliny. *Natural History, Volume IV*. Translated by H. Rackham. Cambridge, Mass.: Harvard University Press, 1986.

Rosenberg, Michael. "The Mother of Invention: Evolutionary Theory, Territoriality, and the Origins of Agriculture." *American Anthropologist* 92 (1990): 399–415.

Solecki, Rose L. "Milling Tools and the Epi-paleolithic in the Near East." *Etudes sur le Quaternaire dans le Monde* 2 (1969): 989–94.

Smith, J. Russell. *Tree Crops: A Permanent Agriculture*. New York: Devin-Adair, 1950.

"White Oak Acorns as Food." *Missouri Botanical Garden Bulletin* 12, no. 2 (1924): 32–33.

Yarnell, Richard Asa. "Aboriginal Relationships between Culture and Plant Life in the Upper Great Lakes Region." Anthropological Papers, Museum of Anthropology, University of Michigan, no. 23 (1964).

THE AGE OF OAK

Anonymous. "Advice to a Norwegian Merchant." In *The Portable Medieval Reader*, edited by James Bruce Ross and Mary Martin McLaughlin. New York: Viking, 1977.

Benson, Ted. *The Timber-Frame Home: Design, Construction, Finishing.* Newtown, Conn.: The Taunton Press, 1988.

Blair, John, and Nigel Ramsay. *English Medieval Industries.* London: The Hambledon Press, 1991.

Bradley, Richard. *An Archaeology of Natural Places.* London: Routledge, 2000.

———. *The Passage of Arms.* Oxford: Oxbow Books, 1998.

Brogger, A. W., and Haakon Shetelig. *The Viking Ships.* London: C. Hurst & Co., 1971.

Bronsted, Johannes. *The Vikings.* New York: Penguin, 1955.

Brough, J. C. S. *Timbers for Woodwork.* New York: Drake, 1969.

Bruce-Mitford, Rupert. *The Sutton Hoo Ship-Burial.* London: British Museum Publications, 1975–83.

Carver, M. O. H., ed. *The Age of Sutton Hoo: The Seventh Century in North-Western Europe.* Woodbridge, UK: Boydell Press, 1992.

Champion, Matthew. *Seahenge: A Contemporary Chronicle.* Norfolk, UK: Barnwell's Timescape Publishing, 2000.

Champion, Timothy, et al. *Prehistoric Europe.* London: Academic Press, 1984.

Chinney, Victor. *Oak Furniture: The British Tradition.* Woodbridge, UK: Baron, 1979.

Christensen, Arne Emil. Author interview.

Coles, Bryony, and John Coles. *Sweet Track to Glastonbury: The Somerset Levels in Prehistory.* London: Thames and Hudson, 1986.

Coles, J. M., et al. "The Use and Character of Wood in Prehistoric Britain and Ireland." *Proceedings of the Prehistoric Society* 44 (1978): 1–45.

Courtenay, Lynn T. "The Westminster Hall Roof and Its 14th-Century Sources." *Journal of the Society of Architectural Historians* 43 (1984): 295–309.

Courtenay, L. T., and R. Mark. "The Westminster Hall Roof: A Historiographic and Structural Study." *Journal of the Society of Architectural Historians* 46 (1987): 374–93.

Crumlin-Pedersen, Ole. Author interview.

Crumlin-Pedersen, Ole, and Birgitte Munch Thye, eds. *The Ship as Symbol in Prehistoric and Medieval Scandinavia.* Copenhagen: National Museum, 1995.

De Oliveira, Manuel Alves, and Leonel De Oliveira. *The Cork.* Lancaster, Penn.: Cork Institute of America, 1995.

The Epic of Gilgamesh. In *The Ancient Near East,* edited by James J. Prichard. Princeton: Princeton University Press, 1975.

Edlin, H. L. *Woodland Crafts in Britain.* London: Batsford, 1949.

Eliade, Mircea. *The Forge and the Crucible.* Chicago: University of Chicago Press, 1978.

Fazio, James R. *The Woodland Steward.* Moscow, Idaho: The Woodland Press, 1985.

Fenwick, Valerie, ed. *The Graveny Boat: A Tenth-Century Find from Kent.* Greenwich: National Maritime Museum, Archaeological Series No. 3. BAR British Series 53 (1978).

Foote, Peter, and David M. Wilson. *The Viking Achievement.* London: Sidgwick and Jackson, 1980.

Gilman, Antonio. "The Development of Social Stratification in Bronze Age Europe." *Current Anthropology* 22, no. 1 (1981): 1–23.

Goriup, Paul, ed. *The New Forest Woodlands*. Oxford: The Forestry Commission, 1991.

Greenhill, Basil. *The Archaeology of Boats and Ships*. Annapolis, Md.: The Naval Institute Press, 1995.

Gustafson, K. H. *The Chemistry of the Tanning Processes*. New York: Academic Press, 1956.

Hansen, Hans Jurgen, ed. *Architecture in Wood*. New York: Viking Press, 1971.

Harding, A. F. *European Societies in the Bronze Age*. Cambridge: Cambridge University Press, 2000.

Harris, Richard. *Discovering Timber-Framed Buildings*. Risborough, UK: Shire, 1999.

Harvey, John. *The Medieval Architect*. New York: St. Martin's Press, 1972.

———. *Medieval Craftsmen*. New York: Drake, 1975.

Haywood, John. *The Historical Atlas of the Vikings*. New York: Penguin, 1995.

Hewett, Cecil A. *English Historic Carpentry*. London: Phillimore, 1980.

Homer. *Iliad*. Translated by Alexander Pope. Glasgow: R. Urie, 1754.

Huang, Yun Sheng, et al. "Westminster Hall's Hammer-Beam Roof: A Technological Reconstruction." *Association for Preservation Technology Bulletin* 20, no. 1 (1988): 8–16.

Kilby, Kenneth. *The Cooper and His Trade*. Fresno, Calif.: Linden, 1971.

Latham, Bryan. *Timber: A Historical Survey of Its Development and Distribution*. London: Harrap, 1957.

Lethwaite, J. G. "Acorns for the Ancestors: The Prehistoric Exploitation of Woodland in the Western Mediterranean." In *Archaeological Aspects of Woodland Ecology*. Oxford: BAR International Series 146 (1982): 217–30.

Linnard, William. "Bark-Stripping in Wales." *Folk Life* 16 (1978): 54–60.

Lubke, Harald. "Submarine Stone Age Settlements as Indicators of Sea-Level Changes and the Coastal Evolution of the Wismar Bay Area." *Greifswalder Geographische Arbeiten* 27 (2002): 203–10.

The Mabinogion. Translated by Gwyn Jones and Thomas Jones. London: J. M. Dent, 1949.

MacDiarmid, Hugh. "In Memoriam: Liam Mac' Ille Iosa." In *Stony Limits and Other Poems.* London: Gollancz, 1936.

McGrail, Sean. *Ancient Boats in Northwest Europe.* **London: Longman, 1998.**

———. *Woodworking Techniques before* A.D. *1500.* Oxford: BAR International Series, 1982.

Morrison, J. S. *Greek and Roman Oared Warships.* London: Oxbow Books, 1996.

Owen, Olwyn, and Magnar Dalland. *Scar: A Viking Boat Burial on Sanday, Orkney.* East Linton, Scotland: Historic Scotland, 1999.

Parsons, James J. "The Acorn-Hog Economy of the Oak Woodlands of Southwestern Spain." *Geographical Review* 52, no. 2 (1962): 211–35.

Pettengell, George. *The Cooper's Craft.* Williamsburg, Va.: The Colonial Williamsburg Foundation, 1967. Videocassette.

The Poetic Edda. Translated by Henry Adams Bellows. New York: Dover, 2004.

Pollard, Joshua. "Inscribing Space: Formal Deposition at the Later Neolithic Monument of Woodhenge, Wiltshire." *Proceedings of the Prehistoric Society* 61 (1995): 137–56.

———. "The Sanctuary, Overton Hill, Wiltshire: A Re-examination." *Proceedings of the Prehistoric Society* 58 (1992): 213–26.

Pryor, Francis. *English Heritage Book of Flag Fen: Prehistoric Fenland Centre.* **London: Batsford, 1991.**

———. *The Flag Fen Basin: Archaeology and Environment of a Fenland Landscape.* London: English Heritage, 2001.

Pyle, Howard. *The Merry Adventures of Robin Hood.* New York: Dover Publications, 1968.

Sawyer, Peter, ed. *The Oxford Illustrated History of the Vikings.* Oxford: Oxford University Press, 1997.

Semenov, S. A. *Prehistoric Technology.* New York: Barnes & Noble, 1964.

Simpson, Jacqueline. *Everyday Life in the Viking Age.* London: Batsford, 1967.

Stevenson, A. C., and R. J. Harrison. "Ancient Forests in Spain: A Model for Land-Use and Dry Forest Management in South-West Spain from 4000 B.C. to 1900 A.D." *Proceedings of the Prehistoric Society* 56 (1992): 227–47.

Sturt, George. *The Wheelwright's Shop.* Cambridge: Cambridge University Press, 1923.

Thompson, J. C. *Manuscript Inks.* Portland, Ore.: The Caber Press, 1996.

Tubbs, Colin R. *The New Forest: An Ecological History.* Newton Abbot: David & Charles, 1968.

Vinsauf, Geoffrey de. *Poetria Nova.* Translated by Margaret F. Nims. Toronto: Pontifical Institute of Medieval Studies, 1967.

Waddell, Gene. "The Design of the Westminster Hall Roof." *Architectural History* 42 (1999): 47–67.

Waterbolk, H. T., and W. Van Zeist. "A Bronze Age Sanctuary in the Raised Bog at Bargeroosterveld." *Helinium* 1 (1961): 5–19.

Waterer, John W. *Leather and Craftsmanship.* London: Faber and Faber, 1950.

———. *Leather in Life, Art and Industry.* London: Faber and Faber, 1952.

Weiss, Harry B., and Grace M. Weiss. *Early Tanning and Currying in New Jersey.* Trenton: New Jersey Agricultural Society, 1959.

Weisstein, Eric W. "Barrel." From *MathWorld*—A Wolfram Web Resource. http://mathworld.wolfram.com/Barrel.html.

Wilson, David M., ed. *Archaeology of Anglo-Saxon England.* Cambridge: Cambridge University Press, 1981.

END OF THE AGE

Albion, R. G. *Forests and Sea Power.* Annapolis, Md.: Naval Institute Press, 1926.

Bamford, Paul W. *Forests and French Sea Power, 1660–1789.* Toronto: University of Toronto Press, 1956.

Boudriot, Jean. *The Seventy Four Gun Ship.* Annapolis, Md.: Naval Institute Press, 1986.

Charnock, John. *An History of Marine Architecture.* London: R. Faulder, 1800–1802.

Dudley, William S., ed. *The Naval War of 1812: A Documentary History.* Washington: Naval Historical Center, 1985.

Glete, Jan. *Navies and Nations: Warships, Navies and State Building in Europe and America, 1500–1860.* Stockholm: Almquist & Wiksell International, 1993.

Gruppe, Henry E. *The Frigates.* Alexandria, Va.: Time-Life Books, 1979.

Holland, A. J. *Ships of British Oak.* Newton Abbot: David & Charles, 1971.

Jane, Fred T. *The British Battle Fleet.* London: S. W. Partridge & Co., 1912.

Lavery, Brian. *The Ship of the Line.* London: Conway Maritime Press, 2003.

Martin, Tyrone G. *Creating a Legend.* Chapel Hill, N.C.: Tryon, 1997.

————. *A Most Fortunate Ship: A Narrative History of "Old Ironsides."* Chester, Conn.: The Globe Pequot Press, 1980.

Norie, J. W. *The Shipwright's Vade Mecum.* London: P. Steele, 1805.

Pepys, Samuel. *King Charles Preserved.* London: Miniature Books, The Rodale Press, 1956.

————. *Memoires Relating to the State of the Royal Navy of England for Ten Years, Determined December 1688.* London: Ben Griffin, 1690.

Robins, F. W. *The Smith: The Traditions and Lore of an Ancient Craft.* London: Rider, 1953.

Smyth, Admiral W. H. *Sailor's Word-Book*. London: Conway Maritime Press, 1991.

Tracy, Nicholas. *Nelson's Battles: The Art of Victory in the Age of Sail*. London: Chatham Publishing, 1996.

Turner, J. M. W. *The Harbours of England*. London: E. Gambart and Co., 1856.

Wood, Virginia Steele. *Live Oaking: Southern Timber for Tall Ships*. Boston: Northeastern University Press, 1981.

OAK ITSELF

Abrahamson, Warren G., et al. "Gall-Inducing Insects Provide Insights into Plant Systematic Relationships." *American Journal of Botany* 85, no. 8 (1998): 1159–65.

Arber, Agnes. *The Natural Philosophy of Plant Form*. Cambridge: Cambridge University Press, 1950.

Anderson, Edgar. "Hybridization of the Habitat." *Evolution* **2, no. 1 (1948): 1–9.**

Axelrod, Daniel I. "Biogeography of Oaks in the Arcto-Tertiary Province." *Annals of the Missouri Botanical Garden* 70 (1983): 629–57.

———. "A Theory of Angiosperm Evolution." *Evolution* 6 (March 1952): 29–60.

Baillie, M. G. L. *A Slice Through Time: Dendrochronology and Precision Dating*. **London: Batsford, 1995.**

Barnett, Raymond J. "The Effect of Burial by Squirrels on Germination and Survival of Oak and Hickory Nuts." *American Midland Naturalist* 98, no. 2 (1977): 319–30.

Bossema, I. "Jays and Oaks: An Eco-ethological Study of Symbiosis." *Behaviour* **70 (1979): 1–117.**

Burger, William C. "The Species Concept in *Quercus*." *Taxon* 24, no. 1 (1975): 45–50.

Collins, Wilkie. *The Woman in White*. Reprint. New York: Bantam, 1985.

Cornell, Howard V. "The Secondary Chemistry and Complex Morphology of Galls Formed by the Cynipinae (Hymenoptera): Why and How." *American Midland Naturalist* 110, no. 2 (1983): 225–33.

Cranwell, Lucy M. "Nothofagus: Living and Fossil." In *Pacific Basin Biogeography: A Symposium*, 1961, edited by J. L. Gressit. Honolulu: Bishop Museum Press, 1963.

Crepet, William L., and Kevin C. Nixon. "Earliest Megafossil Evidence of Fagaceae: Phylogenetic and Biogeographic Implications." *American Journal of Botany* 76, no. 6 (1989): 842–55.

———. "Extinct Transitional Fagaceae from the Oligocene and Their Phylogenetic Implications." *American Journal of Botany* 76, no. 10 (1989): 1493–1505.

Csoka, Gyuri, et al., eds. *The Biology of Gall-Producing Arthropods.* USDA General Technical Report NC-199 (1997).

Darwin, Charles. "Letter to J. D. Hooker." In *More Letters of Charles Darwin*, edited by F. Darwin and A. C. Seward. London: J. Murray, 1903, 7:20.

Delcourt, Paul A., and Hazel R. Delcourt. *Long-Term Forest Dynamics of the Temperate Zone.* London: Springer-Verlag, 1987.

Eiffel, Gustave. In *Le Temps*, Feb. 14, 1887.

Eiffel Tower. Official Web site. www.tour-eiffel.fr.

Flegg, Jim. *Oakwatch: A Seasonal Guide to the Natural History in and around the Oak Tree.* London: Pelham Books, 1985.

Hernandez, V. M., et al. "Ecology of Oak Woodlands in the Sierra Madre Occidental of Mexico." *Journal of the International Oak Society* 4 (1994): 7–15.

Howard, Daniel J., et al. "How Discrete Are Oak Species? Insights from a Hybrid Zone between *Quercus grisea* and *Quercus gambelli*." *Evolution* 5, no. 3 (1997): 747–55.

Hutchins, Ross E. *Galls and Gall Insects.* New York: Dodd, Mead & Co., 1969.

Irgens-Moller, H. "Forest Tree Genetics Research: *Quercus L.*" *Economic Botany* 9, no. 1 (1955): 53–71.

Jensen, Richard J. "Identifying Oaks: The Hybrid Problem." *Journal of the International Oak Society* 6 (1995): 47–54.

Johnson, W. Carter. "The Role of Blue Jays *(Cyanocitta cristata L.)* in the Postglacial Dispersal of Fagaceous Trees in Eastern North America." *Journal of Biogeography* 16, no. 6 (1989): 561–71.

Kuntz, J. E., and A. J. Riker. "Root Grafting in the Translocation of Nutrients and Pathogenic Microorganisms among Forest Trees." In *Nuclear Radiation in Food and Agriculture*, edited by W. Ralph Singleton. New York: D. Van Nostrand, 1956.

Kvacek, Z., and H. Walther. "Paleobotanical Studies in *Fagaceae* of the European Tertiary." *Plant Systematics and Evolution* 182 (1989): 213–29.

Lewington, Richard, and David Streeter. *The Natural History of the Oak Tree.* London: Dorling Kindersley, 1993.

Lewis, C. S. *Mere Christianity.* New York: HarperCollins, 1952.

Mattheck, Claus. *Design in Nature: Learning from Trees.* New York: Springer Verlag, 1998.

———. *Stupsi Explains the Tree.* Karlsruhe, Germany: Karlsruhe Research Centre, 1999.

Mattheck, Claus, and H. Breloer. *The Body Language of Trees.* London: TSO, 1994.

Mattheck, Claus, and Hans Kubler. *Wood: The Internal Optimization of Trees.* New York: Springer Verlag, 1996.

Melville, R. "The Biogeography of *Nothofagus* and *Trigonobalanus* and the Origin of the Fagaceae." *Botanical Journal of the Linnean Society of London* 85 (1982): 75–88.

Miller, Howard, and Samuel Lamb. *The Oaks of North America.* Happy Camp, Calif.: Naturegraph, 1985.

Morgan, Ruth A. *Tree-Ring Studies of Wood Used in Neolithic and Bronze Age Trackways from the Somerset Levels.* Oxford: B.A.R., 1988.

Negi, S. S., and H. B. Naithani. *Oaks of India, Nepal and Bhutan.* Dehra Dun, India: International Book Distributors, 1995.

Nixon, Kevin C. "A Biosystematic Study of *Quercus* Series *Virentes* (The Live Oaks), with Phylogenetic Analyses of Fagales, Fagaceae and Quercus." Dissertation, University of Texas at Austin, 1984.

———. **"Phylogeny and Systematics of the Oaks."** *New York's Food & Life Science Quarterly* **19, no. 1 (1989): 7–10.**

Nixon, Kevin C., and William L. Crepet. "Trigonobalanus (Fagaceae): Taxonomic Status and Phylogenetic Relationships." *American Journal of Botany* 76, no. 6 (1989): 824–41.

Pavlik, Bruce M., et al. *Oaks of California.* Los Olivos, Calif.: Cachuma Press, 1995.

Perry, Thomas O. "The Ecology of Tree Roots and the Practical Significance Thereof." *Journal of Arboriculture* 8, no. 8 (1982): 197–211.

Ramamoorthy, T. P., et al. *Biological Diversity of Mexico: Origins and Distribution.* New York: Oxford University Press, 1993.

Raven, Peter H., and Axelrod, Daniel I. "Angiosperm Biogeography and Past Continental Movements." *Annals of the Missouri Botanical Garden* 61, no. 3 (1974): 540–673.

Rieseberg, Loren H. "The Role of Hybridization in Evolution: Old Wine in New Skins." *American Journal of Botany* 82, no. 7 (1995): 944–53.

Smith, Christopher C. "Food Preferences of Squirrels." *Ecology* 53, no. 1 (1972): 82–91.

Theophrastus. *Enquiry into Plants.* Vol. 3. Edited by A. F. Hort. Cambridge, Mass.: Harvard University Press, 1916.

Thompson, D'Arcy, *On Growth and Form*, Cambridge: Cambridge University Press, 1961.

Tiffney, Bruce H. "The Eocene North Atlantic Land Bridge: Its Importance in Tertiary and Modern Phytogeography of the Northern Hemisphere." *Journal of the Arnold Arobretum* 66 (April 1985): 243–73.

Wolfe, Jack A. "A Paleobotanical Interpretation of Tertiary Climates in the Northern Hemisphere." *American Scientist* 66 (Nov.–Dec. 1978): 694–703.

INDEX

Page numbers in *italics* refer to illustrations.